ENVIRONMENT AND CROP PRODUCTION

Environment and Crop Production

Editors

Ramdane Dris
World Food Ltd., Meri-Rastilantie 3C
00980 Helsinki, Finland

Shri Mohan Jain
Plant Breeding and Genetics Section
Division of Nuclear Techniques in Food and Agriculture
International Atomic Energy Agency, Vienna, Austria

Iqrar A. Khan
Sultan Qaboos University
College of Agriculture
P. O. Box 34, Muscat
Sultanate of Oman

CRC Press
Taylor & Francis Group
Boca Raton London New York

CRC Press is an imprint of the
Taylor & Francis Group, an **informa** business
A SCIENCE PUBLISHERS BOOK

First published 2002 by Science Publishers, Inc.

Published 2010 by CRC Press
Taylor & Francis Group
6000 Broken Sound Parkway NW, Suite
300 Boca Raton, FL 33487-2742

CRC Press is an imprint of Taylor & Francis Group, an Informa business

ISBN-13: 978-1-57808-257-5 (hbk)

Visit the Taylor & Francis Web site at
http://www.taylorandfrancis.com

and the CRC Press Web site at
http://www.crcpress.com

Library of Congress Cataloging-in-Publication Data

Influence of environment on crop production, growth, and disease/editors Ramdane Dris, Iqrar A. Khan, Shri Mohan Jain.
p. cm.
A compilation of articles from a workshop held at the University of Helsinki, Finland, December 2000.
Includes bibliographical references and index.
ISBN 1-57808-257-9
1. Food crops–yields–Congresses. 2. Crop yields–Congresses. 3. Agricultural ecology–Congresses. I. Dris, Ramdane. II. Khan, Iqrar A. (Iqrar Ahmad) III. Jain, Mohan.

SB175 .I53 2002
631.5'58–dc21

2002036504

PREFACE

The book contains papers presented at workshop on the Influence of Environment on Growth, Production, Physiology and Disease of Crops which was held at the University of Helsinki, Finland, December 2000. Articles were selected for inclusion in the book after a process of peer-review and evaluation by an editorial board. The articles deal with various issues related to cultivation technology, crop growth, physiology and disease incidence. Fresh information on current scientific research activities worldwide is also included. The time seemed appropriate for reviewing past and recent achievements and attempting to look into the future. For more details click : www.world-food.net or www.food-farming.info

There is also an urgent need for new approaches or management systems that are multiobjective and integrate 'new' objectives such as quality of produce and cultivation, growing or production methods, quality of the abiotic environment, landscape and nature values into the old objectives. For example, growth of the world fertilizer industry has increasingly been affected by grave concerns, such as environmental pollution and increment in energy consumption, which in many cases have a negative impact on food products and consumer health. In fact, water stress during intensive growth of crops causes inhibition of physiological processes, such as nutrient transport, and hence decreases yield and affects quality performance. For example, limited resources of water in some regions has forced us to seek other sources, such as saline water, for irrigation of crops. Additional benefits of the application of correct fertilizer doses are improved soil protection and environment protection in general and prevention of contamination of soil and groundwaters with nitrites.

The main focus of this book is a review of the environmental factors influencing the growth, development and production of food crops grown under various conditions. Actually, fresh information on their issues is still relatively scarce. The book should be useful to scientists, researchers, students or experts dealing with agronomy, plant physiology, plant nutrition, plant pathology and crop cultivation.

Ramdane Dris

LIST OF REVIEWERS

Ramdane Dris

World Food Ltd., Meri-Rastilantie 3C, 00980, Helsinki, Finland.
E-mail : Dris.uh@Kolumbus.fi
website : www.world-food.net and www.food-farming.info

Iqrar Khan

Sultan Qaboos University, College of Agriculture, P. O. Box 34, PC-123 Muscat, Sultanate of Oman

Raina Niskanen

World Food Ltd., Meri-Rastilantie 3C, 00980, Helsinki, Finland.
website : www.world-food.net and www.food-farming.info

Shri Mohan Jain

International Atomic Energy Agency, Plant Breeding and Genetics Section, Division of Nuclear Techniques in Food and Agriculture, Wagramer Strasse 5, A-1400 Vienna, Austria

Päivi Pyötsiä

University of Helsinki, Department of Applied Biology, P.O. Box 27, FIN-00014, Finland

Juan Pablo Fernández-Trujillo

Universidad Politecnica de Cartagena Paseo Alfonso XIII, Departamento de Ingeniería, Alimentosy del quipamiento Agrícola, ETSIA 30203 Cartagena (Murcia), ESPAÑA

Wei Chin Lin

Agriculture and Agri-Food Canada, Pacific Agri-Food Research Centre, P. O. Box 1000, 6947 #7 Highway, Agassiz, British Columbia. V0M 1A0 Canada

Pawel Wojcik

Research Institute of Pomology and Floriculture Pomologiczną, Department of Fruit Crop Management and Plant Nutrition, 18, 96-100 Skierniewic, POLAND

Ursula Wermund

Projects Manager in Fresh Produce, University of London Wye, Imperial College at Wye, TH Huxley School. Ashford, Kent TN25 5AH

CONTENTS

LIST OF CONTRIBUTORS

K.O. Adenegan

University of Ibadan, Department of Agricultural Economics, Nigeria

A.A. Adegeye

University of Ibadan, Department of Agricultural Economics, Nigeria

O.A. Adekunle

University of Ibadan, Department of Agricultural Extension and Rural Development, Ibadan, Nigeria

L.A. Akinbile

University of Ibadan, Department of Agricultural Extension and Rural Development, Ibadan, Nigeria

Manuel Aluarez

University of Los Llanos "Ezequiel Zamora" (UNELLEZ)

W.M.S.M. Bandara

University of Peradeniya, Department of Crop Science, Faculty of Agriculture, Sri Lanka

Ovidiu Balan

The "Gh.Asachi" Technical University IASI, Facultatea de Mecanica, Bdul.Manageron nr. 63-65, IASI, Cod.6600, Romania

Zita Barkauskiene

Lithuanian Institute of Horticulture, LT - 4335, Babtai, Kaunas distr., Lithuania

Loïc Barthélémy

Unité de Recherche Intégrée en Horticulture Institut National de la Recherche Agronomique Route des Colles, Sophia-Antipolis 06410 Biot France

Alina Basak

Research Institute of Pomology and Floriculture, ul. Pomologiczna 18, 96-100 Skierniewice, Poland.

Philippe Bearez

Unité de Recherche Intégrée en Horticulture Institut National de la Recherche Agronomique Route des Colles, Sophia-Antipolis 06410 Biot France

Pushkar Singh Bora

Department de Tecnologia Quimica e de Alimentos, Unversidade Federal da Paraiba, João Pessoa, Brazil.

Okur Bülent

Ege University, Faculty of Agriculture, Department of Horticulture, 35100 Bornova, Izmir, Turkey

Vasile Bumacov

The Agrarian State University CHISINAU, Republic of Moldova

Jaroslaw Ciura

Agricultural Academy, Department of Vegatable Crops, AI. 29 Listopada 54, 31-425 Cracow, Poland

Vasile Cracium

The "Gh.Asachi" Technical University IASI, Facultatea de Mecanica, Bdul.Manageron nr. 63-65, IASI, Cod.6600, Romania

Z. Cudare

Institute of Biology, University of Latvia, Miera 3, Salaspils LV-2169, Latvia

Maris Daugavietis

Latvian Forest Research Institute 'Silava', Riga Street 111, , Salaspils LV-2169, Latvia

Sezai Delibacak

Ege University, Faculty of Agriculture, Department of Soil Science, 35100 Bornova, Izmir, Turkey

Zeomar Nitão Diniz

Nucleo de Ensino, Pesqiusa e Processamento de alimentos, Universidad Federal de Piauí, Terezina, Brazil

Ramdane Dris

World Food Ltd., Meri-Rastilantie 3C, Helsinki, FIN-00980, Finland www.world-food.net or www.food-farming.info

Sonja Duletic-Lausevic

University of Belgrade, Institute of Botany, Faculty of Biology, Takovska 43, 11000 Belgrade, Yugoslavia

B. Dyki

Research Institute of Vegatable Crops, ul. Konstytucji 3 Maja 1/3, 96-100 Skierniewice, Poland

Maija Eihe

Lavian State Centre of Plant Protection, Latvia

Dursun Eşiyok

Ege University, Faculty of Agriculture, Department of Horticulture, 35100 Bornova, İzmir, Turkey

Ilie Filipescu

Gh. Asachi" Technical University of Iasi, Romania

Zoltan Hegedûs

Szent Istvan University, Institute of Crop Production, H-2100 Gödöllö, Páter Károly u.l., Hungary

André Jaffrin

Unité de Recherche Intégrée en Horticulture Institut National de la Recherche Agronomique Route des Colles, Sophia-Antipolis 06410 Biot France

Tomislav Jemric

Department of Pomology, University of Zagreb, Faculty of Agriculture, Svetosimunska 25, 10 000 Zagreb, Croatia

S. Jha

Indian Institute of Soil Science, Nabibagh, Berasia Road, Bhopal 462038 India

Marton Jolankai

Szent Istvan University, Institute of Crop Production, H-2100 Gödöllö, Páter Károly u.l., Hungary

Shahrokh Khanizadeh

Agriculture and Agri-Food Canada, Horticultural Research Station, 430 Boulevard Gouin, St Jean sur Richelieu, Quebec, Canada J3B 3EB

Krzysztof Klamkowski

Research Institute of Pomology and Floriculture, ul. Pomologiczną, 18, 96-100 Skierniewice, Poland

Dzintra Kreita

Latvia University of Agriculture, Department of Crop Production, Liela iela 2, Jelgava LV-3001, Lativa

I.P. Kozlovskaya

Belarusian state Agricultural Academy, Belarus

N.A.K.G.E. Kumari

University of Peradeniya, Department of Crop Science, Faculty of Agriculture, Sri Lanka

Anda Linina

Latvia University of Agriculture, Department of Crop Production, Liela iela 2, Jelgava LV-3001, Latvia

Iuliana Lupu

Gh. Asachi" Technical University of Iasi, Romania

José López

Institute of Geography, University of Mexico (UNAM)

M.C. Manna

Indian Institute of Soil Science, Nabibagh, Berasia Road, Bhopal 462038 India

Alfonsas Merkys

Institute of Botany Zaliuju ezeru 49, LT-2021 Vilnius, Lithuania

Nkiru T. Meludu

University of Ibadan, Department of Agricultural Extension and Rural Development, Nigeria

Maria Muntanola-Cvetkovic

Institute of Botany, Faculty of Biological Research, Belgrade, Yogoslavia

Ulve Moor

Estonian Agricultural University, Institute of Horticulture, Kreutzwaldi 64, 51014, Tartu, Estonia

Narendra Narain

Department de Tecnologia Química e de Alimentos, Unversidade Federal da Paraiba, João Pessoa, Brazil.

Eftimie Nitescu

Gh. Asachi" Technical University of Iasi, Romania

A.B. Odusina

University of Ibadan, Department of Agricultural Economics, Nigeria

O.I. Oladele

University of Ibadan, Department of Agricultural Extension and Rural Development, Nigeria

F.O. Oladeye

University of Ibadan, Department of Crop Protection and Environmental Biology, Ibadan, Nigeria

M.G. Oludije

University of Ibadan, Department of Agricultural Extension and Rural Development, Ibadan, Nigeria

Nikola Pavicic

Department of Pomology, University of Zagreb, Faculty of Agriculture, Svetosimunska 25, 10000 Zagreb, Croatia

Marija Pecina

Department of Plant Breeding, Genetics and Biometrics, Faculty of Agriculture, University of Zagreb, Svetosimunska 25, 10 000 Zagreb, Croatia

V. Petrova

Institute of Biology, University of Latvia, Miera 3, Salaspils LV-2169, Latvia

Stefan Popescu

Gh. Asachi" Technical University of Iasi, Romania

Malgorzata Poniedzialek

Agricultural Academy, Department of Vegatable Crops, Al. 29 Listopada 54, 31-425 Cracow, Poland

Fiodor Popov

Belarusian SRI of Plant Protection, Republic of Belarus, 223011, Minsk region, p/o Priluki BelSRIPP

Pierre Pyrrha

Unité de Recherche Intégrée en Horticulture Institut National de la Recherche Agronomique Route des Colles, Sophia-Antipolis 06410 Biot France

A.A. Ramalan

Ahmadu Bello University, Agricultural Engineering Department, P.M.B 1044, Zaria, Nigeria

Maria Elena Ruiz

Agrophysics Research Unit, Agricultural University of Havana (UNAH), Apdo. 18, San José de las Lajas, La Habana 32700, Cuba

Antons Ruza

Latvia University of Agriculture, Department of Crop Production, Liela iela 2, Jelgava LV-3001, Latvia

J.K. Saha

Indian Institute of Soil Science, Nabibagh, Berasia Road, Bhopal 462038 India

I. Steinite

Institute of Biology, University of Latvia, Miera 3, Salaspils LV-2169, Latvia

Agnieszka Sekara

Agricultural Academy, Department of Vegatable Crops, Al. 29 Listopada 54, 31-425 Cracow, Poland

Matej Stopar

Agricultural Institute of Slovenia Hacquetova 17, 100 Ljubljana, Slovenia

Zsolt Szentpetery

Szent Istvan University, Institute of Crop Production, H-2100 Gödöllö, Páter Károly u.l., Hungary

E. Szwonek

Research Institute of Pomology and Floriculture, ul. Pomologiczna 18, 96-100 Skierniewice

T.O. Tayo

University of Agriculture, Department of Plant Science and Crop Production, Abeokuta, Nigeria

J. Tenbergs

Institute of Biology, University of Latvia, Miera 3, Salaspils LV-2169, Latvia

A.O. Togun

University of Ibadan, Department of Crop Protection and Environmental Biology, Ibadan, Nigeria

Angel Utset

Agrophysics Research Unit, Agricultural University of Havana (UNAH), Apdo. 18, San José de las Lajas, La Habana 32700, Cuba

Laurent Urban

Unité de Recherche Intégrée en Horticulture Institut National de la Recherche Agronomique Route des Colles, Sophia-Antipolis 06410 Biot France

S. Vabrit

Estonian Agricultural University, Institute of Horticulture, Kreutzwaldi 64, 51014, Tartu Estonia

Alicia del Valle

Agrophysics Research Unit, Agricultural University of Havana (UNAH), Apdo. 18, San José de las Lajas, La Habana 32700, Cuba

Dmitry V. Voitka

Belarussian Scientific-Research Institute of Plant Protection, Minsk region, Belarus 223011

Jelena Vukojevic

University of Belgrade, Institute of Botany, Faculty of Biology, Takovska 43, 11000 Belgrade, Yugoslavia

W.A.P. Weerakkody

University of Peradeniya, Department of Crop Science, Faculty of Agriculture, Sri Lanka

Meilian Yu

Horticultural Research Institute, Shanghai Key Laboratory of Protected Horticulture Technology, Shanghai Academy of Agricultural Sciences, 2901 Beidi Road, Shanghai 201106, China

Ivars Zarins

Institute of Biology, University of Lativa, Miera 3, Salaspils LV-2169, Latvia

Vasil A. Zayats

Uzhgorod National University, Department of Entomology and Agronomy, Biological Faculty, Voloshin str. 46, Uzhgorod 88000, Ukraine

Jingfang Zhang

Faculty of Forest Products and Engineering, Northwest Forestry University, Yanling Shaanxi 712100, P.R. China

Wenbin Zhang

China Agricultural University, 727 Jiaozhou RD, Shanghai, P.R. China

Youlin Zhang

Faculty of Life Science, Shaanxi Normal University, Xi' an Shaanxi 710062, P.R. China

Weimin Zhu

Horticultural Research Institute, Shanghai Key Laboratory of Protected Horticulture Technology, Shanghai Academy of Agricultural Sciences, 2901 Beidi Road, Shanghai 201106, China

Tamara Zherdetskaya

Belarusian SRI of Plant Protection, Republic of Belarus, 223011, Minsk region, p/o Priluki BelSRIPP

1

INFLUENCE OF ORCHARD MANAGEMENT ON GROWTH AND PRODUCTION OF FRUITS

Ramdane Dris
World Food Ltd.
Meri-Rastilantie 3C
Helsinki, FIN-00980
website: www.food-farming.info or www.world-food.net
e-mail: Dris.uh@Kolumbus.fi

The high planting density of apple orchard is used for exploring new efforts to control nutrients and water inputs to maximize yield. Higher plant densities in conjunction with the potential flexibility possible through fertilization with irrigation (fertigation), are resulting in increased attention paid to management of the root zone. Since root volume per plant is lower in high-density planting systems, it is necessary to develop and maintain optimum soil conditions in the root zone. At the same time, a judicious application of nutrients to the aboveground portion of the tree, via foliar sprays, will continue to provide opportunities to optimize fruit quality. The production requirements for orchards, including nutrient additions, will increasingly need to be consistent with society's desire to maintain or improve the quality of soil and water. These ideas will be explored for several important plant nutrients: those which can also adversely affect water quality (nitrogen and phosphorus), those which can have a major impact on fruit quality (calcium and potassium) and some important micronutrients, which are deficient in many apple-growing regions (zinc and boron).

There are numerous and well-recognized factors that may affect the quality of pome fruit at harvest and storage. Generally, fruit quality is associated with climatic and soil conditions, tree characteristics and cultural practices. The apple fruit is especially sensitive to low calcium concentration and susceptible to a number of physiological disorders during storage. The calcium content of apple fruit at maturity is influenced by many cultural and environmental factors. Deficiency of calcium and boron in fruits of apple is a great problem which causes physiological disorders. Transport of boron to fruits in early expansion is reduced by high supply of calcium, while a simultaneous high supply of calcium and nitrate increases both calcium and boron transport into young fruit. There appears to be a barrier for transport of calcium to the fruit, while boron is more freely transported. When dormancy has been broken and the bloom period begins many important factors come into play that can affect the ultimate fruit quality. These relate to pollination, fertilization, seed number and fruit set. The reason why this is important is that the number of seeds may influence the rate of calcium transport to the apples, probably by the production and transport of auxins. On the other hand, fruits which develop from terminal flowers are richer in calcium than those developing from lateral ones. The location of fruits in the tree crown also influences their calcium concentration and the incidence of physiological disorders. Fruits in the upper parts are affected more. Fruits from trees with light crops generally show a lower calcium concentration and are more prone to bitter pit and internal breakdown. It seems that fruit competes with vegetative tissues in some manner for the calcium transported from the root system. Summer pruning can have a strong positive effect on fruit quality, which at least in part is associated with increased calcium levels. Calcium treatments have also been shown to reduce the occurrence of physiological disorders.

It is obvious that in production of fruits the physiological and biochemical role of the various mineral nutrients are often not well considered. As a consequence, uncertainties about the implementation of various recommendations exist. Thus, there is still a great need for applied research on optimal mineral nutrition of fruit trees and also on adequate fertilizer application to make fruit production economically competitive. Qualitative techniques in visual diagnosis are very useful in detecting an individual problem, but by the time visual symptoms are discernible, reduction in yield has already been caused. Therefore, quantitative techniques are preferred. Among them, foliar analysis is one of the most important strategies, since the main metabolic and photosynthetic processes take place in the leaf, and its particular composition reflects the nutritional balance of the whole plant. Analytic criteria consider only a comparison of the leaf concentration for a single element with respect to a reference value. Several limitations arise from this approach that can be partly overcome by use of evaluative nutrient balance methods, such as DRIS (diagnosis and recommendation integrated system). Also, other organs or sap—a good index of plant nutrient contents—can be sampled instead of the leaf.

Optimized mineral nutrition for high yield and good quality often does not need high application of fertilizers. However, self-regulation of nutrient uptake and adaptive mechanisms for nutrient acquisition in fruit trees including rhizosphere processes have to be considered for fertilization recommendation in order to minimize mineral fertilizer use.

2

ABSCISIC ACID CONTENT OF GRAPE DURING STORAGE

Youlin Zhang[1] and Jingfang Zhang[2]
[1]Faculty of Life Science, Shaanxi Normal University,
Xi'an Shaanxi 710062, P. R. China;
[2]Faculty of Forestry Products and Engineering,
Northwest Forestry University, Yanling Shaanxi 712100, P. R. China

INTRODUCTION

Grape (*Vitis* sp.) belongs to the respiratory non-climacteric type of fruits. During storage, ethylene, a plant ripening hormone, was not found to correlate with fruit deterioration while, contrarily, the berry content of abscisic acid (ABA) correlated closely with deterioration of grape clusters and berry drop (Chang and Jacobs, 1973; Wu et al. 1992). Hence it is very important for maintenance of grape freshness during storage to study the change in ABA content in the fruit, the influence of environmental temperature on increment of ABA, and to ascertain suitable environmental conditions and chemicals which would inhibit increase in ABA.

MATERIAL AND METHODS

The Kyoho grape (*Vitis vinifera* L. × *Vitis Labrusca* L.) of Xiwangzhuang village, Baqiao district, Xi'an city was selected for study. The vineyard is situated on a tableland which, due to its higher elevation, has sufficient light and ventilation but does not retain excessive water. The grapevines were eight years old, healthy and neither infected with diseases nor infested by insect pests. The fruit content of soluble solids was over 17%, total acids 0.6-0.9%, sugar to acid ratio 20-25:1 and protopectin ratio to soluble pectin 2.7-2.8:1. Fruit bunches in the middle of the frames were cut at the base of the stem and rotten, small and stiff berries discarded. All experimental samples were harvested simultaneously and the quality of the grape clusters was uniform.

Method of Determining the Content of Internal ABA in Grape Berry

PFB-Br amplication gas chromatography (Huang, 1982; Li et al. 1990) GC-16A electronic gas chromatograph, CR-3A data-processing machine made by Shimadzu (Japan), 1.5% silicone ov-17 2 m × 3 mm glass capillar, and Shimalite W (AW-DM(S)) 60-80 were used. The temperature of the glass capillar was 225°C, gasified temperature 280°C and test temperature 300°C. Nitrogen (N_2) was the carrier gas with flow speed 30 ml min^{-1} and speed of transferring paper was 5 mm min^{-1}. The recycling bottle of KD concentration instrument amplified the back liquid. Chemical agents included PFB-Br made by Sigma Chemical Co. (USA); ABA made by Fluka (USA) and K_2CO_3, and normal hexane and acetone made by Beijing Chemical Factory (China). The purity of N_2 was 99.99%. The retrieval rate of ABA was 97.4% and the minimum quantity tested out was 10^{-3} nbb^{-1}.

Experimental Method of Storage at Room Temperature

Grape-cluster samples were pruned and sprayed with plant-growth regulators and chemicals, then placed in LDPE plastic bags 0.06 mm thick and 20 × 37 cm long and wide. Every bag weighed 500 g and was sealed after filling; a hole was then cut to spread the clusters out. The experimental solution was poured into a small lightproof plastic cup in which absorbent cotton had been placed. A cotton thread was pulled through a hole punched in the cup and secured across the xylem of each cluster with a pin. The solution-soaked thread was left in place for 24 hours, during which the solution was gradually absorbed. The thread was then removed. The lab temperature fluctuated between 29°C and 32°C during grape storage.

Experimental Method of Storage in Mechanical Cold Storage

Grape clusters were treated as follows prior to cold storage:

1) Precooling: Grapes were cooled for 24 h at 0°C prior to storage to release heat absorbed in the field.

2) Fumigation: The cold storage shed was fumigated with 20 g m^{-3} sulphur powder for 24 h. After placement of the fruit bunches in the shed, it was instantly fumigated again for 5 h, then ventilated for 6 h.
3) Treatment: Plant-growth regulators and chemicals were sprayed once a day and after three days the fruits were bagged.
4) Bagging: LDPA plastic bags 0.06 mm thick and 20 × 37 mm long and wide were used. Every bag weighed 500 g and was sealed prior to hanging on the frame.
5) Storage conditions in the cold storage: Temperature 0 ± 0.5°C and relative humidity 90-95%.

Experimental Method of Treatment with Plant-Growth Regulators and Chemicals

The experimental design is shown in Table 2.1.

Table 2.1 Plant-growth regulators, chemicals and their concentrations

Name of reagent	*Code name*	*Concentration (mg^{-1} L^{-1})*
2, 3, 5-triiodobenzoic acid	TIBA	50
6-benzyladenine	6-BA	50
Water (control)	CK	pure water
External abscisic acid	ABA	50
Indole-3-acetic acid	IAA	150
Single berry	Berry	the berry with 5 mm long stalk
Chlorocholine chloride	CCC	
Naphthalene acetic acid	NAA	100
Calcium chloride	$CaCl_2$	200
Ethrel	CEPA	3,000
Gibberellin	GA_3	100
Potassium permanganate	$KMnO_4$	50
Dimethyl aminosuccinamic acid	B_9	saturated solution of $KMnO_4$, 200

RESULTS

1 Change of ABA content during storage of grapes and the influence of different storage temperatures on the formation of ABA.

Untreated grape bunches were stored at low temperature (0 ± 0.5°C) and room temperature (27-32°C). The ABA content of peeled berries was determined at intervals. The curve of ABA content at different temperatures is shown in Figure 2.1.

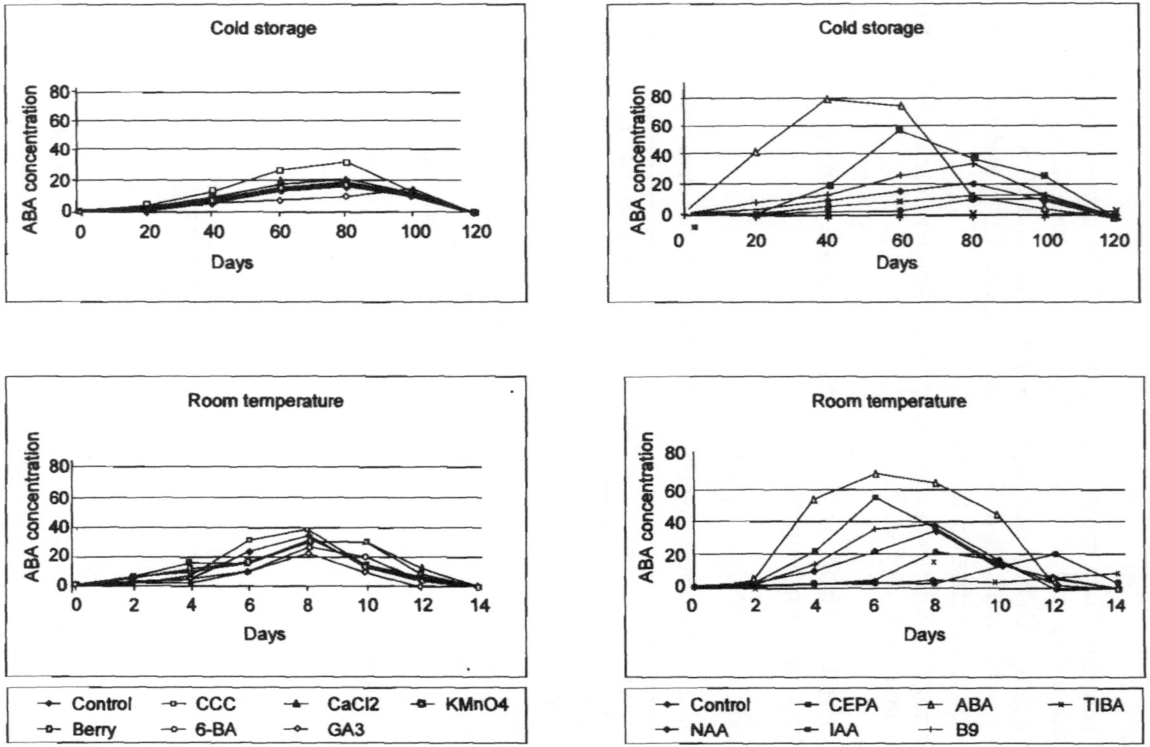

Fig. 2.1 Postharvest ABA concentration (ng g^{-1} Fw) in control and treated grape berries at room temperature (27-32°C) and in cold storage (0 ± 0.5°C)

It can be seen from Figure 2.1 that, depending on storage at room temperature or low temperature, the ABA content in grapes changed from low to high, with a drop characterized by a parabolic curve, with peaks evident in both. The ABA content increased, then declined quickly during storage at room temperature, peaking on the 8th day and falling to near zero (0.85 ng g^{-1} FW) on the 14th day. For grapes stored at low temperature the ABA content changed slowly, peaking (22.45 ng g^{-1} FW) on the 80th day and could not be tested out on the 120th day. This indicates that low temperature could delay the appearance of a peak in ABA and reduce the peak value.

2. Influence of treatment with plant-growth regulators and chemicals on formation of ABA during storage at room temperature.

Grape clusters were treated with nine kinds of plant-growth regulators and two kinds of chemicals, then stored at room temperature (27-32°C). The content of ABA in the grape berries was determined at 2-day intervals and the results shown in Figure 2.1.

Except for treatment with TIBA, the change in ABA content showed peaks, but the time of peaking and peak value differed. Peaking time was reached earlier, at six days of storage, with treatments external ABA and CEPA, with peak values of 68.1 ng g^{-1} FW and 55.53 ng g^{-1} respectively—both higher than in the other treatments. The ABA content then declined over the next six days. IAA, $KMnO_4$ and $CaCl_2$ delayed the appearance of peaking of ABA compared with CK, while treatment with IAA also lowered the peak value. The role of $KMnO_4$ and $CaCl_2$ in diminishing the peak value was not notable. ABA in other treatments peaked on the 8th day, similar to CK. With the four treatments GA_3, NAA, 6-BA and single berry the ABA peak value decreased. For the two treatments CCC and B_9, the peak values were higher than that of CK. The ABA content in grape bunches treated with TIBA always increased gradually over the 14-day storage period, with no peak value evident. The increased quantity of TIBA was 11.44 ng g^{-1} FW, the lowest compared to the peak values of other treatments.

3 Influence of treatments with plant growth regulators and chemicals on the formation of ABA at low temperature.

Grape clusters sprayed with the plant growth regulators and chemicals were stored at a low temperature (0 ± 0.5°C) and the content of ABA in the berries determined at 20-day intervals. The results are shown in Figure 2.1.

The curve of change in ABA content for every treatment during low-temperature storage was similar to the curve obtained at higher temperature Peaking of ABA in grape bunches treated with external

ABA was earliest, i.e., the peak appeared on the 40th day and the value was 81.05 ng g^{-1} FW—the highest among the treatments, followed by 57.42 ng g^{-1} FW on the 60th day of treatment with CEPA. Compared to CK, the three treatments IAA, GA_3 and NAA delayed the peaking time of ABA while CCC and B_9 increased the peak value. ABA change in grape clusters treated with TIBA showed no peak during storage at low temperature, the curve being similar to grape bunches treated with TIBA and stored at room temperature.

DISCUSSION

Organs such as the root, stem, leaf, flower and seed of vascular plants have the ability to synthesize a small quantity of ABA. The ABA contents in the peel of avocado and medallion seed are high, 10 mg g^{-1} and 11.7 mg g^{-1} respectively (Milborrow, 1974). It has been demonstrated that plastid is the main part synthesizing ABA in cells. According to some studies, synthesis of both GA and ABA starts from mevalonic acid and pass the same chemical response—the isoprenoid way. Farnesyl pyrophosphate is the cross-point for formation of ABA and GA (Milborrow, 1974; Loveys and Milborrow, 1984). Treating grape clusters with TIBA may have restrained farnesyl pyrophosphate formation of ABA while concomitantly mediating the path for formation of GA.

The restraining role of GA holds up the formation and expression of functions of ABA. Moreover, TIBA fix IAA (Milborrow, 1974; Loveys and Milborrow, 1984; Ahmad and Magbool, 1992), which could not move in cells indirectly. IAA combined with protein thus remained in the cells in the same state and was not displaced. The quantity of IAA in cells of grape bunches treated with TIBA was large. During long storage periods these fixed states of IAA gradually transited into a free state, thereby maintaining a certain level of IAA in the cells. The content of ABA declined because IAA acted as a restraint.

IAA, GA_3, NAA and 6-BA effectively restrained ABA and hindered the formation of internal ABA in grapes. $CaCl_2$ augmented formation of pectin substances and enhanced their agglutination while $KMnO_4$ was an absorbent of Eth; treatment with these two chemicals had no notable influence on ABA. In the isoprenoid mode of the compound ABA and GA, CCC restrained the kaurene enzyme and GA could not be synthesized while ABA was synthesized. B_9 in large quantity inhibited formation of GA. Since GA restrains formation of ABA, with decline in content of GA, ABA content increases.

CONCLUSIONS

- The change in ABA content in grape during storage is characterized by a parabolic curve with an obvious peak, changing from low to high and then a drop.
- Low temperature (0 ± 0.5°C) is conducive to delaying the appearance of a peak in ABA and reducing the peak value.
- 2,3,5-triiodobenzoic acid (TIBA) greatly restrains formation of ABA. In grapes treated with TIBA and stored at room temperature (27-32°C) for 14 days and at low temperature (0 ± 0.5°C) for 120 days, the content of ABA increased very slowly and no peak appeared.
- Compared to CK, indole-3-abscisic acid (IAA), gibberellin (GA), naphthalene acetic acid (NAA) and benzyladenine (6-BA) can inhibit formation of ABA, delay peaking and reduce the peak value.
- CCC, B_9, CEPA and external ABA enhance formation of ABA and increase the peak value. The ABA peak of grape bunches treated with external ABA and stored at low temperature (0 ± 0.5°C) appeared in 40 days, earlier by 20 days than the peak of CK, (at 60 days), with a peak value 3.2 times that of CK.

References

Abmad, M.J. and Magbool, M. 1992. Chemical changes in grape fruits during maturation and storage. J. Agric. Res. (Pakistan) 30(4): 489-494.

Chang, Y.P. and Jacobs, W.P. 1973. Regulation of abscission and IAA by senescence factor and abscisic acid. Am. J. Bot. 60: 11-16.

Huang, H.B. 1982. Gas chromatograph electronic capture test machine used to determine ABA in plant species. Plant Physiol. Report 1: 64-67.

Li. H. L., Yun, Z. Z. and Yang, Y. P. 1990. X-Br-2,3,4,5 PFBr amplication gas chromatograph used to determine IAA and ABA at the same time. J. Plant 32(3): 205-209.

Loveys, B.R. and Milborrow, B.V. 1984. Metabolism of abscisic acid, pp. 71-104. In: Crozier, A. and Hillman, J. R. (eds) Biosynthesis and Metabolism of Plant Hormones. Cambridge University Press, Cambridge.

Milborrow B.V. 1974. The chemistry and physiology of abscisic acid. Ann. Rev. Plant Physiol., 25: 259-307.

Wu, Y. M., Ren, J. C. and Hua, X. Z. 1992. Drop of grape granule and fresh-keeping storage after picking. J. Plant Physiol.,18(2): 267-272.

3

INFLUENCE OF ENVIRONMENTAL CONDITIONS AT SOWING ON QUALITY OF BROAD-BEAN SEED PRODUCTION

Agnieszka Sekara, Malgorzata Poniedzialek and Jaroslaw Ciura
Agricultural Academy, Department of Vegetable Crops,
Al. 29 Listopada 54, 31-425 Cracow, Poland

INTRODUCTION

Broad-bean seeds (*Vicia faba* L.), also known as common bean, are consumed in Poland at the stage of milk maturity. At this stage their usefulness for consumption is of short duration. Labour-intensive picking and processing of big-seed species has limited the possibility of increasing the cultivation area. Fine-seed species, introduced in recent years, can be processed on lines which are also used for pea processing. Broad-bean seeds for processing should be characterized by proper chemical composition, guaranteeing both high nutritional value and satisfactory organoleptic features of ready-made products (Borowska et al., 1994). Only broad-bean seeds which contain 30-35% dry matter should be used in processing, as this guarantees high nutritional value of the final products. Broad-bean plants are extremely sensitive to water shortage in the soil (Loss and Siddique, 1997). In the climate conditions that prevail in Poland, crops are often exposed to damage caused by spring drought, i.e., during germination and sprouting. Great losses are suffered at harvest when drought occurs in the critical periods of broad-beans, namely the flowering stages and pod formation (Grzesiuk et al., 1997).

This study sets out the correlation between yield of broad-beans, their chemical composition, sowing time and climatic conditions that prevail during growth.

MATERIAL AND METHODS

Between 1998-2000 a field experiment was established in Garlica Murowana, Cracow (Poland) on typical brown soil formed on loess. "Gobik" (PlantiCo Golebiew), a fine-seed variety of broad-bean seed suitable for processing and direct consumption was used in the study. The method of random blocks, size of a single plot 4 m^2, was used, with four replications. Seeds were sown four times in each year: 1998 on 7 April, 24 April, 1 May; 1999–15 April, 26 April, 5 May; 2000–7 April, 17 April and 26 April, at a spacing of 50 × 10 cm. Harvesting was done once, at the milk maturity phase. Following the harvest both general and marketable yields of fresh seeds were calculated. Chemical analyses of the dry matter content included the drying at 70°C, nitrogen determined by Kjeldahl method (converted to protein), sugars by anthrone method and starch by blue index method. ANOVA was applied for statistical analysis and the Student t test (t = 0.05) used.

RESULTS AND DISCUSSION

The yield obtained differed with respect to sowing time (Table 3.1). In 1998, the highest total and marketable yields were obtained from the first two sowing dates. They were significantly higher than the mass of seeds obtained from the third sowing. The first two sowing times were characterized by the highest number of rainfalls, ranging from 250.8 to 273.5 mm, with an average daily temperature of 14°C. In 1999, no significant difference in size of yield from the three sowing periods was found. Due to low rainfall in April and May of that year, seedlings were delayed and irregular. The total yield was low. In 2000, the highest total yield was obtained from the first sowing. However, the quality was poor as the share of marketable yield amounted to 85.7% versus 97.2 and 100% for consecutive sowing times. Studies done by Loss and Siddique (1997) proved that earlier sowing time positively influences the attainment of a high yield. Total yield of the Gobik variety of broad bean, obtained by Wierzbicka (1998) in a comparative study amounted to 8.5 t^{-1} ha^{-1}, with a marketable yield of 7.3 t^{-1} ha^{-1}.

Table 3.1 Total and marketable yield of Gobik variety of broad-bean, cultivated at various times between 1998-2000 (t^{-1} ha^{-1})

Year	*Sowing time*	*Total yield*	*Market yield*
1998	I	7.5	7.3
	II	7.7	7.5
	III	5.0	4.5
	LSD	0.2	0.2
1999	I	1.9	1.8
	II	1.6	1.5
	III	2.3	2.2
	LSD	n.s.	n.s.
2000	I	5.6	4.8
	II	3.6	3.5
	III	1.8	1.8
	LSD	1.2	1.7

n.s : not significant

Table 3.2 Biochemical analysis of the Gobik variety of broad-bean cultivated between 1998 and 2000, at three different sowing times

Year	*Sowing time*	*Dry matter content, %*	*Sugar content, %*	*Starch content, %*	*Protein content, %*
1998	I	21.8	1.6	10.8	16.7
	II	27.0	1.1	21.3	30.8
	III	32.8	1.2	25.8	11.9
	LSD	1.80	0.2	1.8	7.40
1999	I	20.9	1.9	21.7	29.1
	II	26.5	2.0	22.5	28.0
	III	23.3	2.0	22.1	28.9
	LSD	1.70	n.s.	n.s.	0.30
2000	I	20.4	1.8	10.8	—
	II	23.3	2.3	11.4	—
	III	22.3	3.0	11.1	—
	LSD	1.70	0.3	n.s.	—

n.s : not significant

Dry matter content in broad-bean seeds depends on sowing time (Table 3.2). According to Gibczyóski (1995), harvesting of seeds containing 27-35% dry mass gave the highest marketable yield.

In the first year of the experiment, the highest dry mass content was found in seeds harvested in the third sowing period and the lowest one in the first sowing period. A similar tendency was observed in the year 2000, when seeds from the second and third sowing periods contained the most dry mass. In 1999, broad-bean seeds from the second sowing time contained a significantly higher amount of dry mass. Agung and McDonald (1998) stated that early sowing of fine- and medium-size broad-bean in climate conditions similar to those prevailing in Poland, resulted in shortening of the vegetative period, higher yield and higher dry matter content. These results were not been confirmed in our study.

Literature concerning fine-seed varieties of broad-bean contains no complex data referring to chemical composition of seeds harvested during the milk maturity phase. According to Kmiecik and Lisiewska (1994), the content of various elements correlated significantly with the degree of seeds maturity. This study statistically confirmed the correlation between level of dry mass and starch content ($r = 0.6832$ $p < 0.05$).

Sugar content in 1998 was highest in seeds harvested from plants cultivated in the first and third sowing times (Table 3.2). A similar result was observed in 2000, when most sugars were obtained from broad-bean seeds from the third sowing period. However, in 1999, no statistically significant differences were noted. Sugar content in the Gobik variety was similar to the amounts found in large-size varieties, i.e., 2.0-3.2% and depended on the variety (Kmiecik and Lisiewska, 1994).

Between 1999 and 2000 no statistically significant differences were found in the starch content of broad-bean cultivated at different times (Table 3.2). In 1998, the most starch was found in seeds from the third sowing and the least in seeds from the first sowing. According to Grzesiuk and Kulka (1981), prolonged rainfalls led to a decrease in polysaccharides. In the first sowing period in 1998, rainfalls were heaviest and starch content remained at the lowest level.

Table 3.3 Correlation coefficients between rainfalls and temperature and biochemical factors of broad-bean

	Total yield	*Market yield*	*Dry matter*	*Sugars*	*Starch*
Rainfalls	–0.6395	–0.5768	–0.3567	0.6307	–0.1032
Temperature	–0.7030*	–0.7298*	0.1960	0.6552	0.1425

*Market correlations significant at $p < 0.05$

Protein content in broad-bean differed in the 1998 and 1999 cultivations. In the first year highest protein content was noted in seeds from the second sowing time versus the first and third sowing times in the second year. Zadernowski et al. (1994) indicated a negative correlation between total protein content and starch, which was not confirmed by our study.

The correlation coefficient indicated a dependence between temperature at the time of plant vegetation and the yield obtained (Table 3.3). In the case of variety Gobik, an increase in temperature led to a decrease in total and market yield of the seeds.

CONCLUSIONS

- The time of cultivation influenced the total yield of broad-bean variety Gobik. Higher total and marketable crop yields were obtained at the earlier sowing time.
- Broad-bean cultivated in the third sowing period contained the greatest amount of dry mass, sugars and starch.
- A negative correlation between the temperature during the vegetative period and total and marketable yields of broad-bean seeds was observed.

References

Agung, S., and McDonald, G.K. 1998. Effects of seed size and maturity on the growth and yield of faba bean (*Vicia faba* L.). Aust. J. Agric. Res. 49: 79-88.

Borowska, J., Zaderewski, R., Wierzbicka, B. and Roszko, J. 1994. Nowe odmiany bobu—ich wartosc technologiczna i odzywcza. Zeszyty Naukowe ATR w Olsztynie, nr 26: 15-30.

Gibczyóski, P. 1995. Wplyw stopnia dojrzalosci nasion na wartosc odzywcza i przydatnosc technologiczna bobu. Zeszyty Naukowe AR w Krakowie, nr 26: 31-37.

Grzesiuk, S. and Kulka, K. 1981. Fizjologia i biochemia nasion. PWRiL Warszawa

Grzesiuk, S., Filek, W. and Grzesiak, M. 1997. Effect of water stress on the root growth of different drought-tolerant cultivars of legume species. Int. Sci. Mtg. 'Ecophysiological Aspects of Plant Responses to Stress Factors'. Krakow, 12-14 June: 63-69.

Loss, S.P. and Siddique K.H.M. 1997. Adaptation of faba bean to dryland Mediterranean type environments. I. Seed yield and yield components. Field Crop Res. 52 (1-2): 17-28.

Kmiecik, W. and Lisiewska, Z. 1994. Chemical composition of broad bean seeds depending on the cultivar and degree of seed maturity. Folia Hortic. VI/1: 23-30.

Wierzbicka, B. 1998. Przydatnosc nowych drobnonasiennych odmian bobu dla przemys_u przetworczego. Zeszyty Naukowe ATR w Bydgoszczy, nr 215 - Rolnictwo 42: 247-250.

Zadernowski, R., Borowska, J., Markiewicz, K. and Kawecki, Z. 1994. Nowe odmiany bobu—ich wartosc technologiczna i odzywcza. II. Wartosc odzywcza nasion bobu w stadium przydatnosci technologicznej. Acta Academiae Agriculturae AC Technicae Olstenensis 26: 81-88.

4

INFLUENCE OF CALCIUM AND GLYCINEBETAINE ON YIELD AND STORAGE OF POME FRUIT

Ramdane Dris
World Food Ltd.
Meri-Rastilantie 3C
Helsinki, FIN-00980
website: www.food-farming.info or www.world-food.net
e-mail: Dris.uh@Kolumbus.fi

INTRODUCTION

The ability of roots to take calcium can be disturbed by different factors. Only the tips of young roots can take calcium. If the growth of roots is weakened, the uptake of calcium is low. Uptake of calcium differs in different apple varieties because the development of roots is dissimilar. It is important that the moisture condition in soil is good since lack of water decreases the uptake of calcium. If soil pH is low, Al^{3+} ions can disturb calcium uptake. Also, other cations such as NH_4^+, K^+, Mg^{2+}, and Na^+ can compete with calcium in plant uptake (Bangerth, 1974). Leaves are supplied with calcium from the roots through the xylem (Simon, 1978). Once in the leaves, calcium tends to remain there, accumulating as the leaves age. Apple fruits only receive an initial supply of calcium via the xylem because, after the first few weeks of development, transpiration rate falls and later input of nutrient enters via the phloem in which calcium is relatively immobile. Once xylem transport stops, the enlarging fruit receives little further calcium; hence calcium concentration falls with further fruit growth.

The efficiency of transpiration is dependent on the relative humidity of air, wind, temperature and light. Due to the restriction of long-distance transport of calcium to the xylem the transpiration is of special importance for calcium distribution (Marschner, 1974). The relatively low calcium content of fruits compared to the leaves is related to the differences in transpiration rates of these two organs. Consequently a further decrease in the transpiration rate of fruits decreases their calcium content and increases the occurrence of calcium deficiency symptoms. Various parts of the shoot compete for transpiration water and due to their large surface the leaves compete very effectively, presumably withdrawing water from the fruit via the xylem. This competition for water exerts a detrimental effect on the calcium nutrition of the fruit. Because nearly all of the calcium uptake of apples occurs during a period of six weeks after flowering, the calcium content decreases with increase in fruit size (Wilkinson, 1968).

Heavy rainfall and superfluous watering at the end of the growth period can decrease calcium content of apples because their size has increased. There is a tendency that with increasing growth rate of a fruit and therefore increasing intensity of solute influx via the phloem, the calcium content in the fruit decreases (Wiersum, 1966). This relates directly to the reduction of influx of water and calcium via the xylem. Factors which influence the growth rate of fruits such as temperature, presence of other nutrients and leaf/fruit ratio therefore have an indirect effect on the calcium nutrition of fruits (Bangerth and Mostafawi 1969). The tissue nitrogen concentration in apple has an overriding effect on the development of Ca-related metabolic disorders and there are antagonistic relationships between Ca^{2+}, Mg^{2+} and K^{+} (Bangerth, 1974; 1979). Besides the more or less indirect and unspecific regulatory mechanisms for uptake and translocation of calcium, one cannot exclude the existence of specific ones. Various observations as the stimulation of calcium influx into leaves after application of kinetin or the depressing effect of TIBA on calcium influx into fruits such as apple, provide an indication of a possible direct relationship between growth substances and calcium translocation (Shear and Faust, 1970; Marschner, 1974).

PHYSIOLOGICAL ROLE OF CALCIUM

Calcium is present in cell walls and appears to be involved in ion uptake and maintenance of membrane permeability, to have a role in cell division and the assembly of microtubules and in the regulation of many other facets of metabolism (Faust and Shear, 1972; Simon, 1978; Poovaiah et al., 1988). Calcium, in addition to reducing the severity of physiological disorders, alters intracellular and extracellular processes intimately associated with senescence and fruit quality. The rate of softening in apple fruit is sensitive to the calcium status of the flesh (Mason et al., 1974, 1975; Hardenburg and Anderson, 1981). Calcium has a major effect on cell wall structure and membrane integrity. Calcium appears to delay softening in apples by virtue of its ability to delay degradation of the cell wall polymers (Sams and Conway 1984). Calcium appears to serve as an intermolecular binding agent that stabilizes pectin-protein complexes of the middle lamella of the cell wall (Poovaiah et al., 1988). Calcium is not only important in cell wall structure but also has pronounced effects on membrane-associated changes during senescence. Changes in membrane function can markedly alter the respiration climacteric and affect ethylene evolution. Calcium is reported to reduce the rate of ethylene and carbon dioxide evolution in apple fruit (Faust, 1989; Faust and Shear, 1972). Calcium affects the activity of many enzyme systems and metabolic sequences in plant tissues (Wills et al., 1989). Calcium may have a role in the initiation of the normal fruit ripening process. It is also possible that calcium prevents or delays the appearance of physiological disorders by maintaining normal metabolism.

As the symptoms of calcium deficiency occur, fruit tissues are often water-soaked and involve cell breakdown with loss of turgor, which result in internal breakdown (Simon, 1978). Eventually the tissue may become desiccated, yielding a dry, more or less extensive area of necrosis. There is evidence that calcium deficiency renders membranes permeable which would account for loss of turgor and permit cell fluids to invade intercellular spaces. An alternative situation may develop in soft succulent fruits, the cells of which burst under hypotonic conditions. It is suggested that exogenous water may enter in apple through the phloem. Such exogenous water in the intercellular spaces of the fruit may cause cells to swell, thereby cracking the fruit, or may result in bursting of the cells. The low level of calcium impairs the selective permeability of cell membranes and leads to cell injury and necrosis (Sharples, 1971).

Environmental stresses, such as heat, freezing temperatures, salinity, and biotic stresses are known to cause perturbation in cell membranes, cell walls, or both. Therefore, calcium could be expected to play an important role in plant responses to stress. Heat stress results in perturbation of cell membrane transport properties. It can be detected as enhanced ion leakage from the cells. Because extracellular calcium reduces membrane leakage at elevated temperatures, calcium might protect plants from heat stress.

GLYCINEBETAINE IN PLANTS

Glycinebetaine is an organic solute stabilizing protein structures and not inhibiting metabolic processes. It accumulates in the cytoplasm in several species of stressed plants. However, the capacity to synthesize glycinebetaine varies among plant species. One of the basic biological roles of glycinebetaine in plant cells is its effective action as an osmoprotectant. Glycinebetaine has protective effects in plants already in concentration below 1 mM. Accumulation of glycinebetaine leads to osmotic adjustment through mass action, which results in increased water retention and/or sodium exlusion (Ishitani et al., 1995).

Glycinebetaine is a very promising solute to improve drought, heat and salt tolerance of crops and is also a potential compound in the protection of plants against frost and cold injury. Glycinebetaine, in contrast to many salts, is highly compatible with enzymes and thus high glycinebetaine concentrations can be maintained in cells without adverse effects on enzymes and metabolic activity (Palta et al., 1981; Palta and Lee-Stadelmann, 1983).

Even under adequate rainfall or irrigation, plants may experience transient stress during the noon hours of hot days. This transient stress, producing an overshoot of syntoxic protection in the form of abscisic acid (ABA), induces proline accumulation whereas glycinebetaine exerts a negative effect on the increase in proline. Endogenous glycinebetaine has also been associated with maintenance of a significantly higher relative water content and a greater turgor in plant tissues (Saneoka et al., 1995). Thus it might be possible that glycinebetaine relieves the stress response by decreasing the production of ABA.

Two approaches can be applied to introduce glycinebetaine into commercially important crop species or varieties that do not naturally accumulate it or accumulate the compound weakly. One is to insert a complete or an 'improved' glycinebetaine biosynthetic pathway by genetic engineering. Another approach, taken by Cultor Ltd. Finnsugar Bioproducts, is to treat crop plants with exogenous glycinebetaine. To date the responses of over thirty different crops to exogenous glycinebetaine have been studied in several countries. A foliar spray has been the most common procedure employed due to the degradation of glycinebetaine in soil within a few days. Exogenous glycinebetaine penetrates the leaf tissue, which can be enhanced by the addition of adjuvants to the spray solution. Glycinebetaine translocates all over the plant within a few hours and is stable inside the plant tissue, providing systemic protection. It improves photosynthesis and decreases photorespiration in plants grown under salinity and drought (Harinasut et al., 1996). The treatment of tomato plants by glycinebetaine has resulted in yield advantage in Australia and California. In potato, glycinebetaine alleviates scab symptoms whose occurrence is associated with drought and provides protection against freezing injury. In using glycinebetaine, the proper application time is important. When the plant yield is based on fruits (e.g. tomato), application must be done at the onset of flowering. In terms of yield, the most beneficial glycinebetaine dose has been 2 to 3 kg sprayed in water volume 300-500 L ha^{-1}. When individual plants have been treated, the optimum concentration has been 0.05 to 0.1 M (Ishitani et al., 1995; Larher et al., 1996).

References

Bangerth, F. 1974. The function of calcium in the cell and in sub-cellular units of apple fruit. Antagonism between calcium and other elements. Acta Hort. 45: 43-52.

Bangerth, F. 1979. Calcium-related physiological disorders of plants. Ann. Rev. Phytopathol. 17: 97-122.

Bangerth, F. and Mostafawi, M. 1969. Einfluss der Wasserversorgung und des Fruchtgewichtes auf den Mineralstoffgehalt und die Stippigkeit von Apfelfrüchten. Erwerbsobstbau 11:101-104.

Faust, M. 1989. Physiology of Temperate Zone Fruit Trees. John Wiley & Sons, Inc. New York.

Faust, M. and Shear, C. B. 1972. The effect of calcium on respiration of apples. J. Amer. Soc. Hort. Sci. 97: 437-439.

Hardenburg, R. E. and Anderson, R. E. 1981. Keeping qualities of 'Stayman' and 'Delicious' apples treated with calcium chloride, scald inhibitors, and other chemicals. J. Amer. Soc. Hort. Sci. 106: 776-779.

Harinasut, P., Tsutsui, K., Takabe, T., Nomura, M., Takabe, T. and Kishitani, S. 1996. Exogenous glycinebetaine accumulation and increased salt-tolerance in rice seedlings. Biosci. Biochem. 60: 366-68.

Ishitani, M., Nakamura, T., Youn Han, S. and Takabe, T. 1995. Expression of the betaine aldehyde dehydrogenase gene in barley in response to osmotic stress and abscisic acid. Plant Mol. Biol. 27: 307-315.

Larher, F., Rotival-Garnier, N., Lemesle, P., Plasman, M. and Bouchereau, A. 1996. The glycine betaine inhibitory effect on the osmoinduced proline response of rape leaf discs. Plant Sci. 113: 21-31.

Marschner, H. 1974. Calcium nutrition of higher plants. Neth. J. Agric. Sci. 22: 275-282.

Mason, J.L., Drought, B.G. and McDougald, J.M. 1974. Effect of calcium chloride dip on senescent breakdown, firmness and calcium concentration in 'Spartan' apple. Hort. Sci. 9: 596.

Mason, J.L., Jasmin, J.J. and Granger, R.L. 1975. Softening of 'McIntosh' apples reduced by a post-harvest dip in calcium chloride solution plus thickener. Hort. Sci. 10: 524-525.

Palta, J.P., Chen, H.H. and Li., P.H. 1981. Relationship between heat and frost resistance of tuber-bearing *Solanum* species: effect of cold acclimation on heat resistance. Bot. Gaz. 142(3): 311-315.

Palta, J.P. and Lee-Stadelmann, O.Y. 1983. Vacuolated plant cells as ideal osmometer: Reversibility and limits of plasmolysis, and estimation of protoplasm volyme in control and water-stress-tolerant cells. Plant, Cell Environ. 6: 601-610.

Poovaiah, B.W., Glenn, G.M. and Reddy, A.S.N. 1988. Calcium and fruit softening: Physiology and biochemistry. Hort. Rev. 10: 107-152.

Sams, C.E. and Conway, W.S. 1984. Effect of calcium infiltration on ethylene production, respiration rate, soluble polyuronide content, and quality of 'Golden Delicious' apple fruit. J. Amer. Soc. Hort. Sci. 109: 53-57.

Saneoka, H., Nagasaka, C., Hanh, D.T., Yang, W-J., Premachandra, G.S., Joly, R.J. and Rhodes, D. 1995. Salt tolerance of glycinebetaine-deficient and -containing maize lines. Plant Physiol. 107: 631-638.

Sharples, R.O. 1971. Bitter pit of apple. Rep. E. Malling Res. Stn for 1970: 163-167.

Shear, C.B. and Faust, M. 1970. Calcium transport in apple trees. Plant Physiol. 45: 670-674.

Simon, E.W. 1978. The symptoms of calcium deficiency in plants. New Phytol. 80: 1-15.

Wiersum, L.K. 1966. Calcium content of fruits and storage tissue in relation to the mode of water supply. Acta Bot. Neerl. 15: 406-418.

Wilkinson, B.G. 1968. Mineral composition of apples. IX. Uptake of calcium by the fruit. J. Sci. Fd Agric. 19: 646-647.

Wills, R.B.H., McGlasson, W.B., Graham, D., Lee, T.H. and Hall, E.G. 1989. Postharvest. An Introduction to the Physiology and Handling of Fruit and Vegetables. Oxford, Univ. Press, Oxford (3rd ed.) 174 pp.

5

VISCOSITY BEHAVIOUR OF SOME ENZYMATICALLY EXTRACTED TROPICAL FRUIT JUICES

Pushkar Singh Bora[1], Narendra Narain[1] and Zeomar Nitão Diniz[2]
[1]Departamento de Tecnologia Química e de Alimentos, Universidade Federal da Paraiba, Joãa Pessoa, Brazil
e-mail: narain@ct.ufpb.br
[2]Nucleo de Ensino,Pesquisa e Processamento de Alimentos, Universidade Federal de Piauí, Terezina, Brazil

INTRODUCTION

Knowledge of the rheological behaviour of food products is essential for process design and evaluation, quality control and consumer acceptability. Rheological properties of liquid foods have been described in texts on rheology and texture of foodstuffs (Sherman, 1970; Sone, 1972). The rheological properties of a number of tropical fruit purees such as banana (Charm, 1960; Rao et al., 1974), papaya, mango and guava (Rao et al., 1974) and concentrated fruit juices (Rao et al., 1974) have been studied. Improvement in the juice yield and clarity of the papaya, watermelon, cantaloupe, jackfruit, tamarind, grape (Sreenath and Santhanam,1992), maceration of tomato (Pavlov, 1994) and liquefaction of mango pulp (Sreenath and Santhanam, 1992) by treatment with pectinolytic enzymes has been reported by various workers. However, no report is available on the viscosity behaviour of enzymatically extracted fruit juices of different concentrations at varying processing temperatures. The present work was therefore undertaken to furnish data on the viscosity of banana, papaya and mango juices as a function of concentration and temperature.

MATERIAL AND METHODS

Fruits: Ripe banana (*Cavendish musa L.*), papaya (*Carica papaya L.)* and mango (*Mangifera índica* L.) fruits were collected from an orchard located in the municipality of Conde, in the state of Paraíba, Brazil. The fruits were washed in running water and the pulp manually removed with a stainless steel knife and homogenized in a domestic blender at maximum speed.

Enzymatic Extraction of the Juices: Extraction of the fruit juices was carried out by treatment with pectinolytic enzyme Taienzyme-B (Taihei Sangyo Kaiisha Ltd, Japan). Treatment conditions were enzyme concentration 0.075%, temperature 48 ± 2°C and time 2 h. The juices were recovered by centrifugation at 5000 g for 15 min and then concentrated to 20, 30, 40, 50, 60 and 65°Brix by evaporation at atmospheric pressure.

Viscosity Measurement: The viscosity of the juice samples was measured in a Brookfield Synchro-lactric RV model with spindle no.1 throughout the study. A temperature controlled water bath was used to maintain the juice samples at desired temperature. Besides room temperature (28°C), juice samples were maintained at 8, 40, 55 and 70°C and the viscosity measured.

RESULTS AND DISCUSSION

Data on the viscosity of enzymatically extracted banana, mango and papaya juices and their concentrates at different temperatures are presented in Table 5.1. Figure 5.1 also shows the viscosity behaviour of these juices at different degrees of concentrations at room temperature. The viscosity of the banana and papaya juices increased very slowly until a concentration of 40°Brix while the mango juice maintained low viscosity only up to 30°Brix. Increasing the juice concentrations above these concentrations, especially to 65°Brix, had a pronounced effect on their viscosity. In general, at all the juice concentrations mango juice and its concentrates demonstrated higher viscosity than banana and papaya juices and concentrates. Banana juice concentrates until 55°Brix showed lower viscosity than papaya juice concentrates, but above this concentration, there was a pronounced increase in the viscosity of banana juice concentrates. Possibly, the higher viscosities of mango and banana juice concentrates at these higher temperatures could be attributed to the gelatinization of starch in the juices of these fruits. Temperature had a thinning effect on the viscosity of all juice samples. The effect was more pronounced for juices with a higher degree of concentration.

Table 5.1 Effect of juice concentration and temperature on the viscosity of some tropical fruit juices

Fruit juice		*Viscosity (cp) at*					
	Brix	*8°C*	*28°C*	*40°C*	*50°C*	*60°C*	*70°C*
	21.6	3.3	3.0	3.0	2.3	—	2.0
	30.0	3.8	3.3	3.3	2.8	—	2.5
Banana	40.0	15.0	6.0	4.5	3.5	—	3.0
	50.0	20.0	10.5	7.5	6.0	—	4.0
	60.0	57.5	34.8	21.5	13.5	—	9.0
	65.0		283.0	97.5	84.5	—	40.5
	11.6	3.0	3.0	2.5	2.3	—	1.5
	20.0	3.3	3.3	2.6	2.3	—	1.8
	30.0	4.2	3.8	3.3	2.6	—	2.3
Papaya	40.0	7.0	6.8	5.2	3.8	—	3.5
	50.0	22.8	18.0	10.8	6.3	—	5.3
	60.0	43.5	35.0	17.3	12.8	—	8.0
	65.0	54.0	45.0	21.3	16.5	—	9.5
	17.8	8.6	7.0	6.2	6	6	—
	30.0	23.8	9.0	8.6	8.3	8.2	—
Mango	40.0	—	12.8	9.4	8.9	8.4	—
	50.0	68.0	22.0	18.0	14.0	14.0	—
	60.0	188.0	56.0	44.0	30.0	26.0	—
	70.0	720.0	232.0	98.0	74.0	54.0	—

Figure 5.2 shows the effect of shear stress on the shear rate as a function of viscometer dial reading and spindle speed for these juices (at their natural Brix) and their concentrates (55°Brix) at ambient temperature (29 ± 2°C) and at 70°C. In all cases a straight line relationship was observed, indicating the Newtonian behaviour of enzymatically extracted banana, mango and papaya juices and their concentrates. Concentrated fruit juices generally exhibit a pseudoplastic behaviour but a Newtonian behaviour was also observed for depectinized and filtered apple juice in the temperature range of 20-70°C for concentrations varying from 10 to 75° Brix (Saravacos, 1970).

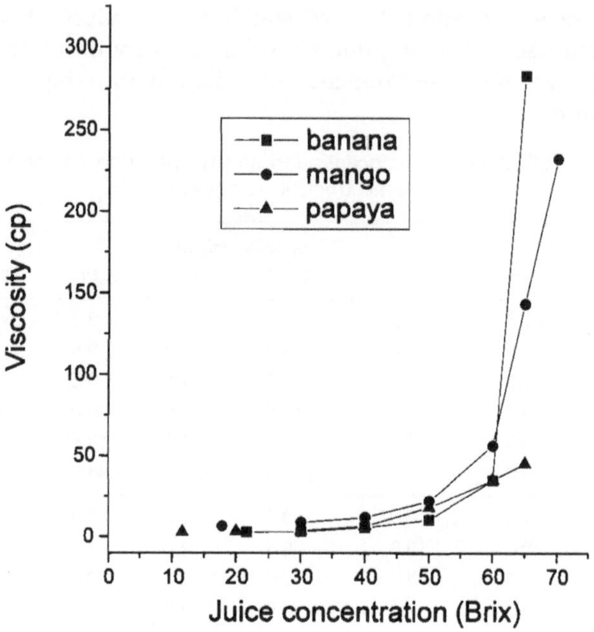

Fig. 5.1 Effect of Brix on viscosity of tropical fruit juices

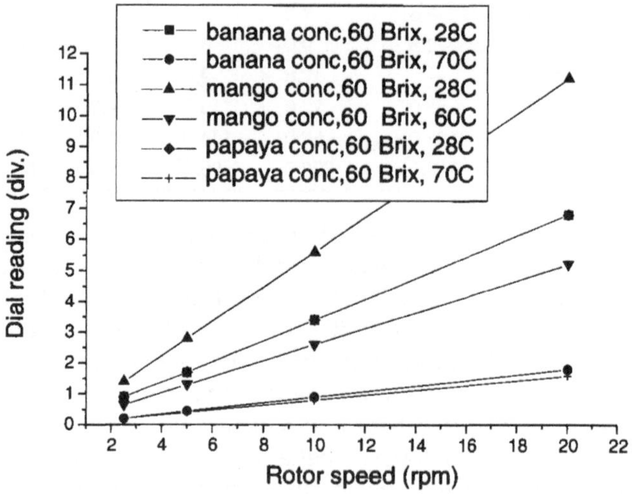

Fig. 5.2 Relationship between stress and shear for different tropical fruit juices (60°Brix) at varying temperatures

References

Charm, S.E. 1960. Viscosity of non-Newtonian food materials. Food Res. 25: 351-362.

Pavlov, N. 1994. Enzymic maceration during tomato processing. Kharanitelna Promishlenost, 43(7): 10-12.

Rao, M.A., Palomino. L N. and Bernhardt, L N. 1974. Flow properties of tropical fruit purees. J. Food Sci. 39: 160-161.

Saravacos, G.D. 1970. Effect of temperature on viscosity of fruit juices and purees. J. Food Sci. 35: 122-125.

Sherman, P. 1970. Industrial Rheology. Academic Press, NY, USA.

Sone, T. 1972. Consistency of Foodstuffs, D. Reidel Publ. Co. Dordrecht, Holland.

Sreenath, H.K and Santhanam, K. 1992. The use of commercial enzyme in white grape juice clarification. J. Ferment. Bioeng. 73(3): 241-243.

6

SUSTAINABLE PRODUCTION AND PLANT GROWTH: INTERACTION BETWEEN CROP YIELD AND QUALITY OF WINTER WHEAT

Antons Ruza, Anda Linina and Dzintra Kreita
Department of Crop Production, Latvia University of Agriculture,
Liela iela 2, Jelgava, Latvia, LV-3001
e-mail: linina@cs.llu.lv

INTRODUCTION

The total harvested wheat production in 2000 in Latvia was 410,300. quintals or 16.6% more compared with total wheat production in 1999. The wheat yield average for 2000 increased by 7.9% and reached 2.6 t ha^{-1} compared to the 1999 cropping year. In 2000, the total area devoted to wheat was 158,000 hectares or 8.2% higher compared to wheat growing areas in 1999.

The production of high-quality wheat grain is a key problem in grain farming in Latvia. Variety selection and best management practices are of great importance in solving the problem of environment (Ruza and Kreita, 1995). The best available technology should be used for fertilizer and pesticide handling and so used as to provide maximum effectiveness with minimal negative side effect on crop and environment (Code for good agricultural practice, 1999). Addition of nitrogen (N) fertilization in accordance with plant requirement is necessary for high yields of winter wheat. To reach the production goal, the soil nitrogen sufficiency level must meet the nutrient demand of a definite crop variety for fertilizer. By taking into account potential yield, the concept allows for a more precise adjustment of the rates of nitrogen fertilizer to be used, but the total doses are split into several applications during the growing season; the length of vegetative period and annual meteorological peculiarities must be considered (Ruza, 1996).

It was found that a supplementary amount of N as top-dressing in the heading stage of winter wheat was mainly used for grain formation, grain size and essentially improving grain quality (Maidl et al., 1998). Spring application of N resulted in lower fertilizer uptake (55%) by plants, but the response of winter wheat to N increased by 35% when applied in later stages of plant growth compared to spring application (Keller et al., 1997).

Grain quality parameters (gluten, protein) were formulated consequently for different factors among which meteorological conditions, genetic peculiarities characteristic of a definite variety and nitrogen fertilizer are the most significant.

To reach the production goal, addition of optimum nitrogen fertilizer rates balanced with fungicide (F) and retardant (R) treatments corresponding to biological peculiarities and yield potential of particular crop varieties is of great importance (Ruza, 1997).

Essential grain yield increase could be achieved with F treatment at stem elongation. Yield increase was mainly attained with an increase in 1000 kernel weight because the treated plant parts (stems, leaves, ears) were protected from senescence (Linina, 2000). Efficiency of retardant depends on plant natural resistance to lodging and meteorological conditions during the growing season, and phytometric status of sowing.

The aim of our studies was to compare the effect of winter wheat varieties with different durations of vegetative period under the influence of nitrogen uptake, and retardant and fungicide treatments on grain yield and baking quality.

MATERIALS AND METHODS

The experiment was carried out at the Research and Training Farm 'Peterlauki' of Latvia University of Agriculture for two years (1998 and 1999). The soil at the site was brown lessive medium loam with medium phosphorus and potassium sufficiency levels, pH_{KCL}-7.0, humus content 2.0%. Conventional measures of agrotechnics were applied in cultivation, NPK (6:26:30) 200 kg ha^{-1}. The seeding rate was 450 treated viable seeds per square metre. During both experimental years the seeding of winter wheat was performed in optimum terms in August.

Plant growth and development depend to a great extent on duration of the vegetative period. The varieties studied could be relatively divided into two groups according to duration of the vegetative period, mid-early: Mironovskaya 808 (Ukrainia), Donskaya polukarlikovaya (Russia), Hugo (Latvia), Krista (Latvia), Sirvintas-1 (Lithuania) and mid-late varieties: Pamyati Fedina (Russia), Sakta (Latvia), Banga (Latvia), Otto (Latvia), Kobra (Poland), Bussard (Germany), Ibis (Germany) and Portal (Germany). In addition, the experimental varieties also differed in terms of geographic origin.

Mid-early varieties received a total dose of mineral nitrogen of 120 kg N ha^{-1} split into two applications. For mid-late varieties, the total dose of mineral nitrogen of 170 kg N ha^{-1} was split into three applications. The initial application was fixed after vegetation started early in spring, the second application at the beginning of stem elongation and the third (for mid-late varieties) at the end of heading. In some treatments, retardant (R) Cykocel (1.5 L ha^{-1}) was applied or the dose was split into 2 applications (R + R) in the following way: two-thirds at tillering and one-third at the beginning of stem elongation. In some variants, particularly in those with repeated incorporation of nitrogen into soil, Tango Super 1.25 L ha^{-1} was applied at inception of heading.

During the years of the experiment distinctive meteorological conditions were not characteristic compared with the long-term average. In 1998, comparatively dry and warm conditions prevailed during the first half of the growth period, while the rainy season lasted from mid-June to end of the harvest period. In conditions of sufficiently high moisture and temperature, rapid vegetative growth was favoured. Frequent gusty wind contributed to lodging and delayed ripening and harvesting of wheat. In 1999 contrarily, the vegetative period was characterized by excessive rainfall and cool weather in spring. Dry and hot conditions prevailed during the following growth period.

All yield parameters were calculated at 14% moisture level and 100% purity. Grain yield and the following baking quality indices were determined: gluten content by *Glutomatic system*, crude protein and Zeleny index by *Infratex*, falling number by *Falling number 1500*, 1000 kernel weight and volume weight according to standard methods used in Latvia. Experiments were analysed statistically by two-factor analysis, regression and correlation.

RESULTS AND DISCUSSION

High nitrogen sufficiency level without F and R treatments allowed high grain yield for mid-early varieties of wheat. A maximum yield of 8.17 t ha^{-1} (2-year average) was obtained with variety Kobra, but in 1998, a grain yield of 8.84 t ha^{-1} was attained under rainfed conditions. With Ibis Kosack and Otto, the grain yields were lower—7.81, 7.53 and 7.03 t ha^{-1} respectively (Fig. 6.1). The resistance of varieties Otto, Kosack and Ibis against diseases was higher.

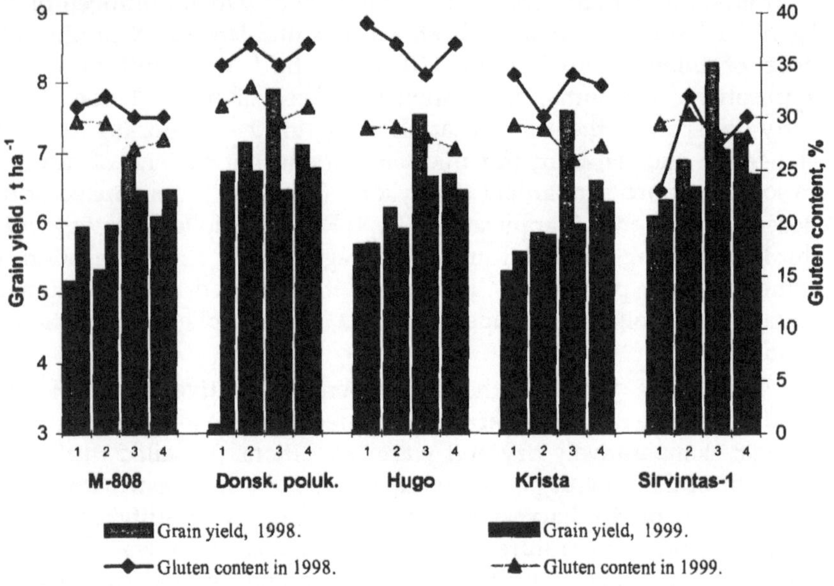

Fig. 6.1 Mid-early winter wheat varieties grain yield t ha^{-1}, gluten content, % 1998 and 1999.

Variants: 1-N_{60+60}, 2-$N_{60+60+F}$, 3-$N_{60+60+R+F}$, 4-$N_{60+60+R+R}$

$LSD_{0.05}$ discrete residuals 0.83 t

$LSD_{0.05}$ varieties = 1.18 t

$LSD_{0.05}$ treated variants = 1.05 t

$LSD_{0.05}$ interaction = 1.40 t

For the aforesaid varieties, spray application of fungicide Tango Super at heading caused reduction in grain yield by 1 to 2% in both experimental years compared to the untreated variant. Grain yield increase was observed with all other wheat varieties. Wheat Kobra responded strongly to fungicide application, resulting in a grain yield increase of 7%, reaching 8.72 t ha^{-1} in the 2-year average.

The F treatments resulted in reduction in grain yield of wheat varieties with higher resistance to diseases as was the case with Ibis, Kosack and Otto when F applications caused plant stress.

Spray application of R Cykocel at tillering provided a yield increase of 5 to 27% in the 2-year average and 6 to 29% under excessive rainfall in 1998. The following wheat varieties responded strongly to spray application of R and the recorded average yield increase was: Otto (27%), Sakta (21%), Bussard (22%), Banga (18%), Kosack (20%) and Portal (16%).

Split R application based on the growth stage (at tillering and beginning of stem elongation) resulted in yield reduction by 8 to 18% in 1998 for all the tested varieties. Grain yields for wheat were characterized by high gluten content. Exceeding 30% for particular crop varieties, these grains corresponded to gluten quality group I or II. Owing to particular weather conditions (severe drought in the second half of the vegetative period in 1999), gluten content in grain ranged from 26.0 to 32.9% and corresponded to gluten quality group III. Therefore the grain was not fit for human consumption. Such low quality gluten was observed with varieties Sakta, Banga, Otto, Kosack, Kobra and Bussard. With mid-early varieties, mean grain yield of 6.42 t ha^{-1} was highest with variety Donskaja polukarlikovaja,

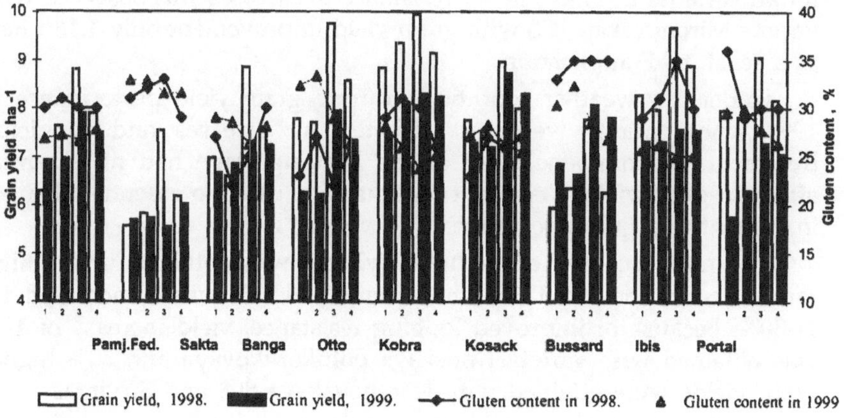

Fig. 6.2 Mid-late winter wheat varieties grain yield t ha^{-1}, gluten content, % 1998 and 1999.

Variants: 1-$N_{60+70+40}$, 2-$N_{60+70+40+F}$, 3-$N_{60+70+40+R+F}$, 4-$N_{60+70+40+R+R+F}$

$LSD_{0.05}$ discrete residuals 0.78 t

$LSD_{0.05}$ varieties = 1.09 t

$LSD_{0.05}$ treated variants = 0.72 t

$LSD_{0.05}$ interaction = 0.96 t

surpassing that of Mironovskaja 808 by 16% (Fig. 6.2). Essential yield increase was obtained with Sirvintas-1, with grain yield average at 5.6 t ha^{-1}.

Excessive rainfall and lower mean temperature during the period of vegetation occurred in 1998 compared to 1999. Nevertheless, the produced mean grain yield was only 7% lower in 1998 for all varieties tested.

Grain gluten content was high in both 1998 and 1999, ranging from 29 to 30%, except for Sirvintas-1 which showed a sharp decline, reaching only 23% in 1998 under conditions of excessive rainfall during yield formation. The rest of the varieties tested showed high grain gluten content corresponding to gluten quality groups I and II in 1998 under high rainfall conditions as well.

Greater average maximum temperatures and rainfall deficiency in 1999 were the factors that lowered grain moisture content to the critical limit, resulting in hindered gluten quality formation for variety Donskaja polukarlikovaja. Corresponding to gluten quality group III, the grain of this wheat variety was not fit for human consumption as for the rest of the winter wheat varieties, the grain gluten content corresponded to quality groups I and II. Fungicide Tango Super spray applied to wheat crop post-anthesis, provided yield increase of 3-17% for all varieties in 1998. Higher resistance to diseases was observed with variety Mironovskaja 808 with grain yield improvement only 1.16 t ha^{-1} as a result of F application.

Under dry weather conditions, during grain yield performance in 1999 wheat plants were less affected by diseases and fungicide treatments at commencement of the heading stage had no essential effect on grain yield. Retardant treatments caused no essential changes in content and quality of gluten.

The most positive effect of R Cykocel was obtained under high rainfall conditions in 1998 when grain yield of wheat increased by 11-30%. Because of improved lodging resistance, yield increase of 11% was obtained with variety Donskaya polukarlikovaya and 30% higher grain yields were attained with Mironovskaya 808 and Sirvintas-1.

The 2- year averages show that mean grain yield was greatest with R Cykocel for both varieties with lower and those with good lodging resistance.

Under high rainfall conditions, split R treatments in mid-early wheat sowings resulted in reduction of grain yields of winter wheat varieties by 10-13%, and increased gluten content in grain without essential changes in gluten quality group.

Falling number is essentially affected by enzymic activity in grain. The more active the grain, the lower the falling number. With falling number exceeding 220 s, wheat grain meets the requirements for food grain. Falling number was high for all tested wheat varieties (241 to 430 s), except Banga (200 s).

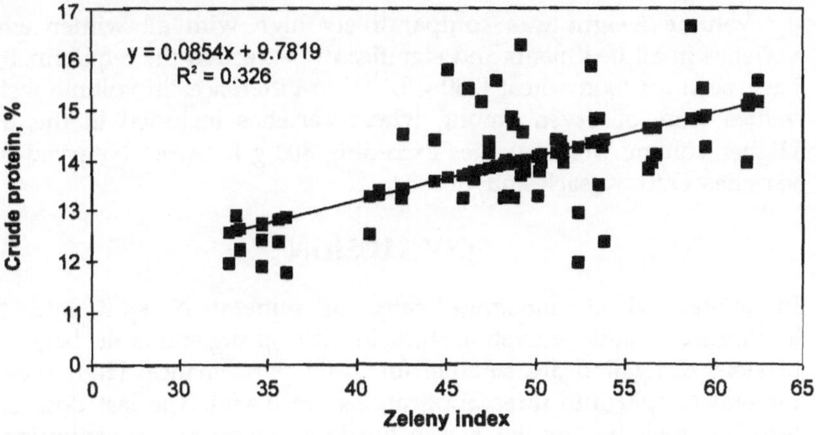

Fig. 6.3 Correlation between crude protein (%) and Zeleny index in 1998.

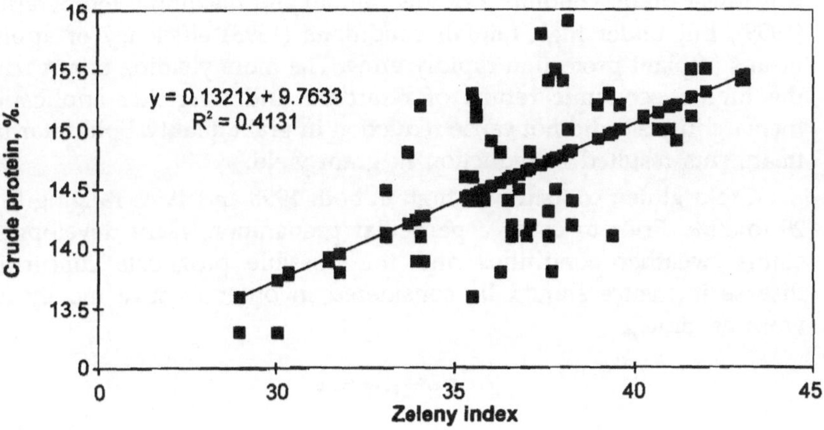

Fig. 6.4 Correlation between crude protein (%) and Zeleny index in 1999.

Crude protein (CP) changed from 11.8 to 16.5% while the Zeleny index ranged from 28-63 s. These grains were useful for direct baking or mixing with comparatively weak flour.

Close correlation (Fig. 6.3 and Fig. 6.4) was found between the CP and Zeleny index in 1998—r = 0.571 (n = 56 and $\alpha_{0.05}$ = 0.253) and 1999—r = 0.643.

Volume weight was comparatively high with all winter wheat varieties in all treatments and significantly exceeded the requirement of basic seed for food wheat (730 g L^{-1}). No differences in volume weight values were observed among wheat varieties included in the trial. Higher volume weight values exceeding 800 g L^{-1} were observed with varieties Otto, Kosack and Busard.

CONCLUSION

In winter wheat, moderate rates of mineral N split into two applications—after vegetation starts in early spring and at the beginning of stem elongation are suitable for grain performance. High rates of mineral N split into three applications are useful. The last dose of N fertilizer may be applied at the heading stage, thus contributing to higher quality grain yield.

In winter wheat, the efficiency of fungicide and retardant treatments was lower under conditions of low rainfall and maximum temperatures (1999), but under high rainfall conditions (1998) efficiency of applied means of plant protection rapidly grew. The more yielding crop variety, the higher economic return of retardant and fungicide application. Increase in yield did not cause reduction in grain quality. Split retardant treatments resulted in reduction in grain yield.

Grain gluten content was high in both 1998 and 1999, ranging from 29 to 30%. Prior to using a particular preparation, plant development status, weather conditions and the possible prognosis concerning disease incidence should be considered in order to save money and working time.

References

Code for good agricultural practice. 1999. Jelgava LV, 103 pp.

Maidl, F., Sticksel, E., Retzer, F. and Fischbeck, G. 1998. Effect of varied N-fertilisation on yield formation of winter wheat under particular consideration of main stems and trillers. J. Agron. Crop. Sci. 180, (1): 15-22.

Keller, E., Hanus H. and Heyland, K. 1997. Grundlagen der Landwirtschaftlicher Pflanzenproduktion. Ulmen, 860 S.

Linina, A. 2000. The influence of different agrochemical, on winter wheat seed yield and seed quality. pp. 28-33. International Scientific Conference Reports: Science Latvia Europe. Jelgava, LV.

Ruza, A. 1996. Agroecological Substantiation of Wheat and Barley Yields and Their Quality. Jelgava, LV 75 p.

Ruza, A. 1997. Interactions between components of crop ability, pp. 203-212. In : Problems of Field Crop Husbandry and Soil Management in Baltic States. Kaunas, LT.

Ruza, A. and Kreita, Dz. 1995. Effect of nitrogen fertilizer on winter wheat yield and quality pp. 104-108 In: Problems of Field Crop Husbandry and Soil Management in Baltic States, no. 182. Tartu, EE.

7

APPLE FLOWER AND FRUIT THINNING

Matej Stopar
Agricultural Institute of Slovenia Hacquetova 17, 1000 Ljubljana, Solvenia
e-mail: matej.stopar@kis-h2.si

INTRODUCTION

Apple trees frequently produce far too many flowers and fruits to obtain regular marketable crops throughout the year. Most apple cultivars will retain this heavy set of fruit throughout the growing season, resulting in small, poorly coloured, low-quality fruit. Early and ample thinning is necessary for adequate effects on fruit size, colour, and return bloom (Bergh, 1990, 1992). Thinning solely by hand is often impossible because of labour cost, vulnerability and poor effect on return bloom.

THINNING APPLICATION

Fruitlet drop after application of thinning compounds is a complicated phenomenon and not very predictable. The most consistent thinner for apples is carbaryl. Since it is not very climate dependent it can be used from petal fall to mid-June (Wertheim, 1998). The drawbacks of carbaryl are its high toxicity for bees and negative influence on pest predators in the orchard (Hislop and Prokopy, 1981). So, carbaryl is not permitted in integrated fruit production and will disappear from the European market in a few years. Other thinning agents used on the basis of their hormonal action are much kinder to the environment. The most used compounds of plant-growth regulators with active thinner ingredients are: 1-naphthaleneacetic acid (NAA), naphthaleneacetamide (NAAm), ethephon (ethylene inducer) and benzyladenine (BA). According to the standards of WHO, these growth regulators (NAA, NAAm and ethephon) are classified as plant protection products with minimum toxicity. Their active ingredients are systematized in Products unlikely to present acute hazard in normal use' (Agrochemical Handbook, 1994; Pesticide Manual, 1995). Another reason why these products belong to the class of less toxic compounds is the strength of their action. These active ingredients are always applied as a dilute spray and in very small concentrations. NAA, NAAm and ethephon thin in concentrations up to 10 ppm, 50 ppm or 300 ppm respectively. They are allowed in integrated production in most countries. BA is a natural plant compound and acts as a thinner with concentrations up to 100 ppm.

The main problem of all 'hormonal' thinning agents is the inconsistency of their action. Many environmental and tree factors affect the outcome (Williams, 1979, 1994; Forshey, 1987). NAA and NAAm cause problems with their varying influence on thinning from year to year. Their activity is ecologically dependent and overthinning has been observed in some years. In addition, some authors have reported that fruit growth was not sufficiently enhanced and referred to thinning intensity in the crown (Thomann and Botzner, 1996; Black et al., 1995). Thinning of 'Red Delicious' with NAA could cause severe fruit growth inhibition known as 'pygmy' fruit (Greene, 1943; Bukovac et al., 1994). Recently, a synthetic cytokinin 6-benzyladenine (BA) was found to be a good apple thinning agent (Wismer et al., 1995; Greene et al., 1990). In addition and independent of its thinning action, BA enhances growth of apple fruit (Elfving and Cline, 1993; Greene, 1993). Ethephon induces ethylene biosynthesis and thereby promotes maturation of the fruitlet abscission zone layer (Horton, 1984; Sexton et al., 1989). Ethephon has a good effect on return bloom but the thinning action is very temperature dependent (Grausland, 1988). In some years ethephon can cause overthinning (Link, 1986).

In recent years the European Working Group for Fruit thinning has done some research work on new environment-friendly thinning compounds with consistent performance. One of the best alternative compounds is ammonium thiosulphate (ATS) which could be registered as a fertilizer. ATS applied at 1% spraying concentration has low toxicity for the environment and shows promise on apple trees. The concentration burns the stigmas of apple flowers and is only slightly phytotoxic for leaves. Two weeks after spraying 0.5% – 1.5% ATS there is no visible phytotoxicity in the crown while the fruit set is diminished properly, depending on cultivars used (Stopar, 2000).

Because of the high environmental dependency of all the above-mentioned thinning agents, experiments are necessary to find the thinner with the best efficiency in a given environment.

References

Agrochemical Handbook, The 1994. The Royal Society of Chemistry, Thomas Graham House, Cambridge, England.

Bergh, O. 1990. Effect of time of hand-thinning on apple fruit size. S. Afr. J. Plant Soil 7: 1-10.

Bergh, O. 1992. Cumulative effect of time of hand-thinning on fruit size of Golden Delicious and Grany Smith apples. S. Afr. Plant Soil 9: 64-67.

Black, B.L., Bukovac, M.J. and Hull, J. 1995. Effect of spray volume and time of NAA application on fruit size and cropping of Redchief 'Delicious' apple. Sci. Hortic. 64: 253-264.

Bukovac, M.J., Black, B.L., Hull, J. and Stopar, M. 1994. Interaction between NAA and BA on cropping and fruit size in 'Delicious' and 'Empire' apples. Hort. Sci. 29: 472.

Elfving, D.C. and Cline, R.A. 1993. Benzyladenine and other chemicals for thinning 'Empire' apple trees. J.Amer. Soc. Hort. Sci. 118 (5): 593-598.

Forshey, C.G. 1987. A review of chemical thinning of apples. New England Fruit Meetings 93: 68-73.

Grauslund, J. 1988. Fruit thinning VI. Further experiments on chemical thinning of the apple cultivar 'Summerred'. Danish J. Plant Soil Sci., 92: 269-273.

Greene, D.W. 1993. A review of the use of benzyladenine (BA) as a chemical thinner for apples. Acta Hortic., 329: 231-236.

Greene, D.W., Autio, W.R. and Miller, P. 1990. Thinning activity of benzyladenine on several apple cultivars. J. Amer. Soc. Hort. Sci. 115 (3): 394-400.

Greene, L. 1943. Growth regulators and fruit set with Starking apples. Proc. Amer. Soc. Hort. Sci. 42: 149-150.

Hislop, R.G. and Prokopy, R.J. 1981. Integrated management of phytophagous mites in Massachusetts (U.S.A.) apple orchard. 2. Influence of pesticides on the predator *Amblyseious fallacis* (Acarina: Phytoseiidae) under laboratory and field conditions. Prot. Ecol. 3: 157-172.

Horton, R.F. 1984. Carbon Dioxide Flux and Ethylene Production in Leaves. V: Ethylene and Plant Development (J.A. Robert in G.A. Tucker). Butterworths, London, pp. 37-45.

Link, H. 1986. Growth regulators. Acta Hortic. 179: 207-213.

Pesticide Manual, The 1995. The Royal Society of Chemistry, Thomas Graham House, Cambridge, England.

Sexton, R., Lewis, L.N., Trewavas, A.J. and Kelly, P. 1989. Ethylene and abscission. V: Manipulation of fruiting (C.J. Wright). Butterworths, London, pp. 173-193.

Stopar, M. 2000. Testing of new fruit-thinning compounds. Sodobno kmetijstvo 33: 51-54.

Thomann, N. and Botzner, B. 1996. Sind Spritzungen mit NAA für die Wachstumsregulierung geeignet? Obstbau Weinbau 33 (11): 304-306.

Wertheim, S.J. 1998. Chemical thinning of deciduous fruit trees. Acta Hortic. 463: 445-462.

Williams, M.W. 1979. Chemical thinning of apples. Hort. Rev. 1: 270-300.

Williams, M.W. 1994. Factors influencing chemical thinning and update on new chemical thinning agents. Compact Fruit Tree 27: 115-122.

Wismer, P.T., Proctor, J.T.A. and Elfving, D.C. 1995. Benzyladenine affects cell division and cell size during apple fruit thinning. J. Amer. Soc. Hort. Sci. 120 (5): 802-807.

8

EFFECTS OF CHEMICAL THINNERS ON APPLE

Nikola Pavicic[1], Marija Pecina[2] and Tomislav Jemric[3]
[1,3]Department of Pomology
[2]Department of Plant Breeding, Genetics and Biometrics Faculty of Agriculture, University of Zagreb, Svetosimunska 25, 10 000 Zegreb, Croatia
e-mail: tjemric@agr.hr

INTRODUCTION

Fruit thinning can improve fruit size and annual bearing of apple trees. Plant-growth regulators and some insecticides are used for thinning. Results from recently published literature point to advantages of combining certain growth regulators and insecticides for thinning. A combination of carbaryl and NAA has given good results in some years but caused excessive thinning in others (Rogers and Williams, 1977). Generally, mixtures of growth regulators (Elfving and Cline, 1993a), or growth regulators and insecticides (Byers and Carbaugh, 1991; Marini, 1996) are used, particularly for cultivars not easily thinned (Byers and Carbaugh, 1991). In such cases, it is possible to reduce the concentration of an individual thinner. In contrast to positive effects, such as increased fruit mass (Pavicic and Paulic, 1989; Bound et al., 1993) and reduction in biennial bearing (Link, 1978), chemical thinning of fruit may sometimes cause adverse effects, such as reduced yield (Schumacher and Fanhkrauser, 1972; Rogers and Williams, 1977; Byers and Carbaugh, 1991; Elfving and Cline, 1993a, b; McArtney et al., 1995; Marini, 1996), fruit russeting (Pavicic and Paulic, 1989; Byers and Carbaugh, 1991;

Bound et al., 1993), fruit deformation (Rogers and Williams, 1977), poor fruit colour (Byers and Carbaugh, 1991; Link, 1991) and lower calcium concentrations in fruit (Elfving and Cline, 1993a). These days the issue of the possible unfavourable effect of chemical thinners on the environment has become important. This fact led us to check whether it is possible to decrease the dose of thinner by combining it with another. The present study evaluates the effects of three combinations of thinners applied on 'Idared' apple in doses lower by 50% than usual.

MATERIAL AND METHODS

The orchard was planted in 1981 near Donja Zelina and the training form was spindle-bush on M.9 rootstock. It was situated at 45'58" N and 16'15" E and located 155 m above sea level.

The experiment was designed as randomized complete block design with four replications, with each plot (3.8 m × 1 m) consisting of three trees. Herbicides were applied to the grass between rows. Thinning treatments were applied when fruit diameter averaged 10.5 mm (1990), 10.0 mm (1991) and 11.1 mm (1992). The same three thinner combinations (Table 8.1) were used, with water as with control. The same trees were given the same treatment during all three years. In all combinations the spray volume was 1100 L ha^{-1} and no adjuvant was added.

Table 8.1 Doses and combinations of chemical thinners applied in 'Idared' orchard, 1990-1992

Designation	*Combination*	*Dose (mg L^{-1})*
EC	Ethephon (48%) + Carbaryl (50% 1-naphthyl methyl carbamate)	330 + 2200
DM	Dirigol (50% NAD) + Mesurol (50% 4-methyl thio-3,5-xylyl methyl carbamate)	73 + 500
EM	Ethephon (48% ethephon) + Mesurol (50% 4-methyl thio-3,5-xylyl methyl carbamate)	360 + 500

The following traits were measured: yield (kg $tree^{-1}$), fruit mass (g) and number of fruits per tree. Fruit coloration was evaluated visually as poorly coloured (< 50% of fruit surface red) and well coloured (> 50% of fruit surface red). For every tratment 100 fruits were randomly taken across replications and evaluated. Air temperature data was obtained from the State Weather Bureau station at Maksimir, about 20 km from the orchard. The sum of heat units was calculated using a base temperature of 7°C (De Jong and Goudriaan, 1989) and the formula:

GDD = $(t_{max} + t_{min})/2 - 7$, where t_{max} is absolute daily maximum and t_{min} daily absolute minimum air temperature. Trial data for yield, fruit mass and number of fruits per tree were analysed using general linear model procedure with SAS System 6.12 software (SAS Institute Inc., 1989). The data were analysed annually and combined over years. Tukey's honest significant difference (HSD) test with studentized range for mean comparisons was used. Coloration data was analysed using logistic regression available in SAS and thinner combinations were tested against control for significance using the chi-square test.

RESULTS AND DISCUSSION

Annual analyses showed no significant difference in yield between thinned and unthinned trees in any year (Table 8.2). When Mesurol or Carbaryl were applied as single thinners, the same effect was observed (Crnko and Marn, 1988). However, there was significant increase in fruit mass with the application of DM in 1991 and 1992, EC in 1990 and 1992, and EM in 1992, and decrease in number of fruits per tree with the application of EC in 1990 and 1992.

The combined data over years showed no significant difference between treatment yield, but mean yield in 1991 was significantly lower than in 1990. Fruit mass was significantly higher in 1991 and 1992 compared to the starting year 1990, while the opposite was true for number of fruits per tree. All three thinner combinations had greater average fruit mass than control. Slight increase in fruit mass was noticed with the EC combination. Such an outcome could be explained by the positive effect of carbaryl on fruit mass (McArtney et al., 1995; Wismer et al., 1995; Marini, 1996), while ethephon behaved neutrally (Elfving and Cline, 1993a) or had a positive effect (Jones et. al., 1997). The effect of carbaryl on fruit mass may be explained by its effect on cell size (Wismer et al., 1995). In our research this effect was only partial because the number of fruits was also significantly reduced, especially in 1992 when thinning with EC caused nearly 50% reduction compared to control. In addition, EC was the only thinner which decreased the number in general.

Therefore, increase in fruit mass is also a result of reduced fruit number because of lower competition for assimilates. The increase in fruit mass with application of DM in 1991 and 1992 was a direct effect, since for these years the number of fruits was not significantly reduced. Mesurol sometimes shows such a tendency but the effect is not significant (Crnko and Marn, 1988). Dirigol is a growth regulator that increases fruit size (Schumacher and Fanhkrauser, 1972). The positive

Table 8.2 Results of SAS *glm* procedure for yield, fruit mass and number of fruits per tree in thinning experiments of 'Idared' apple (MSD = minimal significant difference. Means in a row followed by the same letter do no differ significantly on the basis of Tukey's HSD test on P < 0.05)

		Treatment									Year Mean	R = range	
Variable	*Year*	*DM*		*EC*		*EM*		*Control*		*Pr > F*		*p < 0.05*	*Pr > F*
Yield	1990	12.67	n.s.	10.58	n.s.	12.75	n.s.	14.34	n.s.	0.269	12.58	a	
(kg•tree^{-1})	1991	11.84	n.s.	9.17	n.s.	9.92	n.s.	9.33	n.s.	0.742	10.06	b	
	1992	10.58	n.s.	8.42	n.s.	11.34	n.s.	13.00	n.s.	0.329	10.83	ab	*0.042*
	Treat. Mean	11.69	n.s.	9.39	n.s.	11.33	n.s.	12.22	n.s.	0.107			
	Year x Treat.									0.762			
										MSD (Year) p < 0.05 2.372 p < 0.01 3.260			
Fruit mass	1990	180.00	ab	200.00	a	170.00	ab	160.00	b	0.015	177.50	b	
(g)	1991	240.00	a	230.00	ab	220.00	ab	180.00	b	0.051	217.50	a	
	1992	230.00	a	250.00	a	240.00	a	180.00	b	0.002	225.00	a	*0.002*
	Treat. Mean	216.67	a	226.67	a	210.00	a	173.33	b	0.010			
	Year x Treat.									0.193			
	MSD (Treat.) p < 0.05 36.184 p < 0.01 51.984									MSD (Year) p < 0.05 26.731 p < 0.01 36.747			
Number of fruits	1990	72.59	ab	52.91	b	75.52	ab	89.59	a	0.046	72.66	b	
(per tree)	1991	49.70	n.s.	39.30	n.s.	44.92	n.s.	51.66	n.s.	0.671	46.40	a	
	1992	45.90	ab	33.93	b	48.33	ab	72.08	a	0.012	50.06	a	*0.002*
	Treat. Mean	56.06	ab	42.05	b	56.26	ab	71.11	a	0.011			
	Year x Treat.									0.514			
	MSD (Treat.) p < 0.05 19.093 p < 0.01 27.430									MSD (Year) p < 0.05 15.517 p < 0.01 21.330			

n.s.: not significant

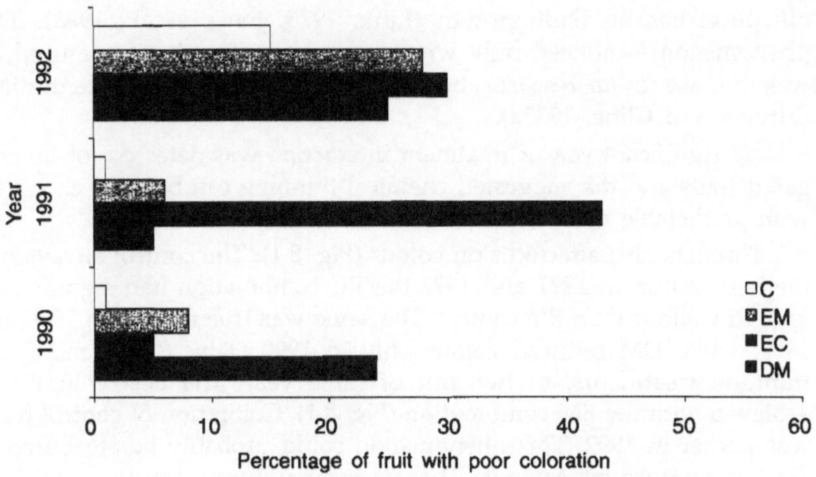

C = Control; EM = Ethephon + Mesurol;
EC = Ethephon + Carbaryl; DM = Dirigol + Measurol

Fig. 8.1 Effect of thinning treatments on fruit coloration intensity.

Table 8.3 P-values of chi-square test for fruit coloration data (comparison between control and thinning treatments)

		Year	
	1990	*1991*	*1992*
Cont. vs. EC	0.1352	0.0001	0.0002
DM	0.0009	0.1352	0.0798
EM	0.0443	0.0907	0.0272

Table 8.4 Sums of growing–degree–days (GDD) between full bloom and harvest, and number of days with temperature above 30°C in July and August

Year	*GDD (full bloom - harvest)*	*No. of days with temp. above 30°C*	
		July	*August*
1990	1846.4	6	9
1991	1755.0	3	3
1992	1984.7	6	23

effect of DM on fruit mass could be due to the interaction of NAD and Mesurol or even of NAD itself. The EM combination produced the smallest, although nonsignificant, fruit among thinner treatments in 1990 and 1991. Such an outcome could be explained by the effect ethephon has on fruit growth (Link, 1978; Jones et al., 1983). This phenomenon is noticed only when higher concentrations are used, as was the case in our research, but was not found at lower concentrations (Elfving and Cline, 1993a).

No significant year × treatment interaction was detected for investigated traits and the suggested chemical thinning can be used each year with predictable results.

Thinners also affected skin colour (Fig. 8.1). The control always had the best colour. In 1991 and 1992 the EC combination had significantly less skin colour than the control. The same was true for EM in 1990 and 1992, while DM reduced colour only in 1990 (Table 8.3). Among the thinning treatments, in two out of three years the best colour was achieved with the EM combination (Fig. 8.1). Coloration of control fruits was poorer in 1992. This phenomenon could probably be attributed to air temperature in August. The air temperature data for 1990-1992 (Table 8.4) showed that the GDD sum was almost equal in 1990 and 1991 from full bloom to harvest, while in 1992 it was higher. There were 23 days with temperature above 30°C during August 1992, which may have reduced fruit colour.

The referenced literature points to the possible impact of other factors in addition to climate on colour development (Byers and Carbaugh, 1991). Because two key enzymes in anthocyanin synthesis are phenylalanine ammonia lyase (PAL; EC 4.3.1.5) (Leyva et al., 1995) and flavanone synthase, we tried to interrelate the effect of temperature with the activity of these enzymes. In lettuce, PAL activity is reduced at higher temperatures (Loaizaverde et al., 1997). In apple, pigment synthesis correlated more closely with flavanone synthase activity than PAL (Faragher and Chalmers, 1977). For Cox's Orange, there are two periods of anthocyanin synthesis during fruit growth (Saure, 1990), the first during the period between the second and eighth week after full bloom, and the second between the twelfth and twenty-second week, i.e., prior to harvest. The period of high air temperatures corresponds with the second anthocyanin synthesis period, which indicates possible blockage in activity of PAL and/or flavanone synthase and poorer production of colour. The period before harvest is extremely important in anthocyanin synthesis because its level increases three times during this period (Knee, 1972).

The effect of the thinner on colour is not sufficiently described in the literature. McArtney and coauthors (1995) reported that chemical thinners do not adversely affect fruit coloration, but Byers and Carbaugh (1991) and Link (1991) stated the opposite.

CONCLUSION

Lower doses of growth regulators in combination with insecticides could be effectively used for fruit thinning. The present paper shows that thinners do not affect yield but do increase fruit mass and reduce fruit colour. In addition to thinning, fruit coloration may also be affected by air temperatures before harvest.

References

Bound, S. A., Jones, K.M., Graham, B., Oakford, M.J. and Tichon, M. 1993. Modelling the effects of timing and rates of application of benzyladenine as a secondary thinner of Fuji apple after etephon. J. Hort. Sci. 68: 967-973.

Byers, R. E. and Carbaugh, D.H. 1991. Effect of chemical thinning sprays on apple fruit set. Hort Tech., pp. 41-48.

Crnko, J. and Marn, M. 1988. Biennial results of fruit thinning of apple cultivars Jonagold and Gloster with Mesurol WP 50. Jugosl. voc. 22: 393-398. (in Serbian)

De Jong, T.M. and Goudriaan, J. 1989. Modeling peach fruit growth and carbohydrate requirements: Revaluation of the double-sigmoid growth pattern J. Amer. Hort. Sci. 114: 800-804.

Elfving, D.C. and Cline, R.A. 1993a. Cytokinin and etephon affect crop load, shoot growth, and nutrient concentration of Empire apple trees. HortSci. 28: 1011-1014.

Elfving, D.C. and Cline, R.A. 1993b. Benzyladenine and other chemicals for thinning Empire apple trees. J. Amer. Soc. Hort. Sci. 118: 593-598.

Faragher, J.D. and Chalmers, D.J. 1977. Regulation of anthocyanin synthesis in apple skin. III. Involvement of phenylalanine ammonia lyase. Austr. J. Plant Physiol. 4: 133-141.

Jones, K.M., Koen, T.B. and Meredith, R.J. 1983. Thinning Golden Delicious apples using etephon sprays. J. Hort. Sci. 58: 381-388.

Jones, K.M., Bound, S.A., Summers, C.R. and Oakford, M.J. 1997. Preliminary examination of thinning strategies on young Jonagold and Pink Lady apples. Austral. J. Exp. Agr. 37: 377-382.

Knee, J. 1972. Anthocyanin, carotenoid and chlorophyll changes in the peel of Cox's Orange Pippin apples during ripening on and off the tree. J. Exp. Bot. 23: 184-196.

Leyva A., Jarrilo, J. A., Salinas, J. and Martinezzapater, J.M. 1995. Low temperature induces the accumulation of phenylalanine ammonia-lyase and chalcone synthase MRNAs of *Arabidopsis thaliana* in a light-dependent manner. Plant Physiol. 108: 39-46.

Link, H. 1978. Gegenwärtiger Stand der Fruchtausdünnung. Erwerbsobstbau 5: 88-91.

Link, H. 1991. Die aktuelle Situation bei der Fruchtausdünnung. Obst und Garten 5: 240-244.

Loaizaverde, J.G., Tomasbarbera, F.A. and Saltveit, M.E. 1997. Effect of intensity and duration of heat-shock treatments on wound-induced phenolic metabolism in iceberg lettuce. J. Amer. Soc. Hort. Sci. 122: 873-877.

Marini, R. P. 1996. Chemically thinning spur Delicious apples with carbaryl, NAA, and etephon at various stages of fruit development. HortTech. 6: 241-246.

McArtney, S.J., Tustin, D.S., Seymour, S., Cashmore, W. and Looney, N.E. 1995. Benzyladenine and carbaryl effects on fruit thinning and the enhancement of return flowering of three apple cultivars. J. Hort. Sci. 70: 287-296.

Pavicic, N. and Paulic, N. 1989. Effect of chemical thinning on the goldspur (Auvil spur) cultivar on the amount of yield, diameter, rustiness and the degree of fruit ripeness. Agron. Glas. 1-2: 3-11. (in Croatian)

Rogers, B.L. and Williams, G. R. 1977. Chemical thinning of spur-type Delicious apple fruit. Virginia Fruit 65: 23-28.

SAS Institute, Inc. 1989. SAS/STAT® User's Guide, version 6, 4th ed., vol. 1 (943 pp.) and 2 (846 pp.). SAS Institute Inc., Cary, N.C.

Saure, M.C. 1990. External control of anthocyanin formation in apple. Sci. Hort. 42: 181-218.

Schumacher, R. and Fanhkrauser, F. 1972. Chemische Fruchtausdünnung mit Sevin und Naphtylacetamid in kombinierter Anwendung. Schweiz. Zeitschr. f. Obst-und Weinbau 11: 264-268

Wismer, P.T., Proctor, J.T.A. and Elfving, D.C. 1995. Benzyladenine affects cell division and cell size during apple fruit thinning. J. Amer. Soc. Hort. Sci. 120: 802-805.

9

EVALUATION OF FLOWERING CRAB-APPLES AS POLLINIZERS OF COMMERCIAL APPLE CULTIVARS

Shakrokh Khanizadeh
Agriculture and Agri-Food Canada, Horticultural Research Station,
430 Boulevard Gouin, St Jean sur Richelieu, Quebec, Canada J3B 3E6
e-mail: khanizadehs@EM.AGR.CA

INTRODUCTION

Most diploid varieties of apples are partly or wholly self-incompatible, setting little or no fruit with their own pollen. They must be cross-pollinated with pollen from another variety before a satisfactory crop can be obtained. McIntosh apple cultivar has been reported to be slightly self-fruitful (Westwood, 1993) and, as such, needs to be cross-pollinated. Flowering crab-apples have been described as effective pollinizers of commercial apple trees (Crassweller et al., 1980). The advantages of using these ornamental trees are their prolific source of flowers, pollen, and fruit, which are obviously distinct from the main variety. This eliminates the difficulty of keeping varieties separate at harvest time. Crab-apple trees are best used as pollinizers for high-density treewall or hedgerow orchards of a single variety (Westwood, 1993). They can be planted between trees of the main variety and pruned to a narrow pillar to prevent crowding of other trees and to reduce shading.

Plants to be used for cross-pollination must flower regularly every year, produce a high percentage of viable pollen, and be compatible with the main variety. The pollinizer should start shedding its pollen before anther dehiscence of the main variety rather than after to increase the efficiency of pollination.

This report evaluates the potential use of 19 varieties of crab-apples as pollinizers of McIntosh/M.26 apples. Flowering season and pollen compatibility, measured by number of fruits produced, as well as number of seeds per fruit were determined.

MATERIAL AND METHODS

Pollen was extracted from 18 crab apple-cultivars and Liberty, a scab-resistant cultivar, and refrigerated at 4°C until further use. Concomitantly, preselected branches of McIntosh/M.26 trees were covered with cheesecloth bags. Branches were bagged a few days before flower opening to prevent unwanted cross-pollination by bees. After the flowers had fully opened, the bags were carefully removed for a few minutes and 100 blossoms per tree hand-pollinated with the pollen extracted from the test cultivars, using a sterilized brush. A comparable branch with 100 blossoms was also tagged in each tree and used as an indicator of fruit set under natural conditions. At harvest time, the number of apples produced by hand-pollinated and open-pollinated branches, as well as number of seeds produced by these apples, were counted. Blooming dates of crab-apple cultivars and Liberty were recorded and checked for overlap with the main fruiting commercial variety, McIntosh. Flower colours of the different pollinizers were also noted. The experiment was performed in 1990 and repeated in 1991.

Microscopic observation. Some of the extracted pollen was mixed with a 10% sucrose solution and its germination monitored microscopically over a 48-hour period.

Statistical analysis. In order to detect differences between pollinizers, per cent fruit number was transformed following an arc sine transformation.

RESULTS AND DISCUSSION

On a 2-year average, the blooming season of McIntosh trees was about 8 days. Forty per cent of the flowers were open around mid-May and 60% petal fall was observed around the 22nd. The bloom season of a pollinizer should show an overlap with the main fruiting commercial variety if it is to play its role effectivly. Moreover, and because of the coolness of the early bloom season resulting in a poor honeybee flight, thus poor pollination, it is more important to provide an abundant

Table 9.1 Ability of crab-apple pollens or Liberty cultivar to pollinize McIntosh trees as measured by the arc sine √% fruit number

Number of apples hand-pollinized with indicated source of pollen	*Source of pollen*	*Number of apples obtained through natural self-pollination*	*F-value*	*Flower Colour*
31 ± 8	Aldenhamensis	30 ± 8	0.07 NS	P
32 ± 4	Almey	30 ± 7	0.18 NS	R
39 ± 6	Carmine	22 ± 1	26.07*	W
35 ± 7	Dolgo	20 ± 5	63.35*	W
38 ± 2	Eleyi Neville Copeman	22 ± 1	610.78*	R
37 ± 4	Evereste	30 ± 2	3.90 NS	W
37 ± 1	Hopa	16 ± 2	491.89**	R
46 ± 12	Hupehensis	28 ± 8	2.39 NS	R
21 ± 8	Kobenza	16 ± 6	15.24 NS	R
40 ± 9	Liberty	20 ± 7	20.37*	W
34 ± 1	Makamik	26 ± 3	22.31*	R
32 ± 11	Malus Peach	29 ± 5	0.32 NS	R
37 ± 4	Pioneer Scarlett	22 ± 6	16.16*	R
36 ± 8	Radiant	21 ± 7	182.14**	R
14 ± 9	Soulardi Red Tip	18 ± 4	1.27 NS	W
32 ± 6	Sun Dog	21 ± 4	49.92*	R
39 ± 12	V, C, 2 crab	26 ± 10	9.82 NS	R
46 ± 13	Van Esseltine	22 ± 2	8.51 NS	P
27 ± 6	Zofia	17 ± 6	11.69 NS	R
34 ± 1	Overall average	23 ± 1	84.6**	

*statistically significant at the 0.05 level.

**statistically significant at the 0.01 level.

Legend: NS-Not significant. R-red, P-pink, W-white.

supply of pollen early in the bloom season rather than later (Way, 1978). Liberty and 11 crab-apple cultivars did not show 100% overlap with the McIntosh blooming period. Kobenza and Aldenhamensis had shorter blooming seasons than McIntosh, showing 40% flower opening 1 day later and 60% petal fall 2 days earlier than McIntosh. Hupehensis and Hopa blooming season started at the same time as McIntosh, which means that the amount of pollen available to the bees was not abundant early in the McIntosh blooming season. Moreover, Hupehensis is a triploid variety, and these are known to produce little good viable pollen (Way, 1978).

The number of harvested apples was highest when Hopa or Radiant were used as a pollinizer (Table 9.1). Carmine, Dolgo, Eleyi Neville

Copeman, Liberty, Makamik, Pioneer Scarlett and Sun Dog also resulted in a significantly higher number of fruits than did branches that were left to cross-pollinate naturally (Table 9.1). This was not reflected in per cent pollen germination, as Van Esseline shows the highest rate of pollen germination but does not show a significant increase in number of harvested apples (Tables 9.1 and 9.2). This is somehow expected, since it is not only the quality of the pollen that plays an important role in the success of pollination, but also the compatibility of this pollen with the main commercial variety. Among the cultivars resulting in a high number of harvested apples, only Carmine, Makamik and Pioneer Scarlett show a bloom season overlap with McIntosh.

Table 9.2 Per cent pollen germination as determined microscopically after mixing the pollen with a 10% sucrose solution

	% pollen germination after 48 hours	
	in 1990	*in 1991*
Aldenhamensis	4.3	3.35
Almey	14.9	3.88
Carmine	2.2	3.36
Dolgo	3.3	3.75
Eleyi Neville Cop.	2.1	2.9
Evereste	8.2	2.23
Hopa	8.1	4.9
Hupehensis	0.8	3.49
Kobenza	6.2	2.8
Liberty	2.5	2.86
Makamik	11.9	2.76
Malus Peach	12.2	2.63
Pioneer Scarlett	13.2	4.55
Radiant	8.3	1.87
Soulardi Red Tip	1.9	5.12
Sun Dog	7.6	4.09
VC2-Crab	13.8	5.14
Van Esseltine	16.8	3.67
Zofia	7.1	4.17

In England, honeybee interchange between crab-apple cultivars and commercial varieties was reported to be highest when flowers were of the same colour (Kendall and Smith, 1975). It would be important to confirm these findings under our conditions. If such is the case, then Carmine, Dolgo, Evereste, Liberty and Soulardi Red Tip would be better pollinizers for McIntosh since their flowers have a colour coinciding with that of McIntosh.

It was found that the number of seeds in an apple variety determines somehow its storage capacities and that the minimum required for a good storage consisted of 6 seeds per apple. Soulardi Red Tip and Carmine showed an average seed number lower than 6 (Table 9.3). However, Carmine had a high number of seeds per apple in 1990 and the low number of seeds in 1991 is the result of missing data.

Table 9.3 Average number of seeds produced in 'McIntosh' apples when different pollinizer sources were used

	Number of seeds per apple	
	in 1990[1]	*in 1991*[2]
Aldenhamensis	8	6
Almey	8	5
Carmine	9	1*
Dolgo	9	6
Eleyi Neville Cop.	8	5
Evereste	8	8
Hopa	9	6
Hupehensis	8	4*
Kobenza	8	7
Liberty	8	6
Makamik	8	6
Malus Peach	8	7*
Pioneer Scarlett	7	4*
Radiant	9	7
Soulardi Red Tip	2	2*
Sun Dog	8	5
VC2-Crab	8	3
Van Esseltine	9	4
Zofia	8	7

[1]Average number of seeds in 20 apples.
[2]Average number of seeds in 15 apples.
*Denotes missing data.

As a conclusion, it would appear that planting different pollinizers in 'McIntosh' orchards would be best. In addition to shedding abundant pollen at different times, the trees would show different degrees of susceptibility to pests, thus making a source of pollen continuously available. This study indicates that the combination of Dolgo, Carmine and Liberty as white flowering varieties would appear to be suitable as a pollinizer source for McIntosh. Dolgo would be the initial pollen shedder and therefore would provide abundant pollen for McIntosh flowers early in the flowering season. Carmine and Liberty would contribute their pollen later, since Carmine flowers one day earlier and Liberty 4 days later than McIntosh. Among the red flowering varieties, Eleyi Neville Copeman, Makamik, Pioneer Scarlett and Hopa would be good candidates as a mixture pollinizer source for McIntosh. Makamik, Liberty and Dolgo were reported to be resistant to apple scab and should be tested under local conditions. Hopa, on the other hand, was reported to be susceptible to apple scab in Ohio climatic conditions, and should also be tested under local conditions (Smith and Treaster, 1986). Dolgo was reported to produce large apples that could be mixed with apples from the main commercial variety, and should also be checked before further use (Way, 1978).

It appears, from this experiment, that crab-apples have a good potential for use as pollinizers for 'McIntosh'. Further experimentation is needed before reliable recommendations can be made.

References

Crassweller, R.M., Ferree, D.C. and Nichols, L.P. 1980. Flowering crab apples as potential pollinizers for commercial apple cultivars. J. Amer. Soc. Hort. Sci. 105: 475-477.

Kendall, D. A. and Smith, B. D. 1975. The foraging behaviour of honeybees on ornamental *Malus* spp. used as pollinizers in apple orchards. Appl. Ecol. 12: 464-471.

Smith, E. M. and Treaster, S. A. 1986. Evaluation of flowering crabapple susceptibility to apple scab in Ohio-1986. Ohio Agr. Res. and Dev. Ctr. Res. Circ. Ornamental Plants-1987: a Summary of Research, pp. 3-7.

Way, R. 1978. Crab apple pollinizers. Proc. 123rd Annual Mtg: NY State Hort. Soc. 123: 97-101.

Westwood, M.N.1993. Temperate-zone Pomology: Physiology and Culture (3rd ed.). Timber Press, Inc. Portland, USA.

10

RESPONSES OF CROP PLANTS TO LOW, NON-FREEZING TEMPERATURES

Krzysztof Klamkowski
Research Institute of Pomology and Floriculture,
ul. Pomologiczna 18, 96-100 Skierniewice, Poland
e-mail: kklamk@insad.pl

Chilling causes disturbances in the course of many vital processes in crop plants, which may result in growth inhibition, reduction in yield, or deterioration in quality. The unfavourable effects of chilling are essentially caused by changes taking place at the cell level, which modify the progress of many physiological processes. The present work is a review of the available literature on the influence of chilling on these processes.

Under natural conditions, plants, both wild and those cultivated by man, are often exposed to unfavourable environmental factors. The influence of these factors may cause growth inhibition, disturbances in generative development, and numerous metabolic changes which may cause permanent injuries to cells, tissues and organs, or even result in an individual's death. To survive the effects of unfavourable factors, plants, upon receiving a stimulus from the environment, initiate defensive responses in order to protect cellular structures and maintain their function.

A serious problem in the cultivation of warm-season plants is chilling, i.e., temperatures in the range 0 – 15°C. In the initial phase of the vegetative period a serious threat to plants is posed by prolonged cold spring weather often occurring in central, eastern and northern Europe. The level of thermal tolerance which allows uninterrupted progress of vital processes of plants is established genetically and determined by specific properties of enzymes with varying sensitivity to changing temperatures and by the stability of cellular structures (mainly membranes), which ensures continuity of metabolic reactions.

Temperature affects all of the vital processes in plants. It has a modifying effect on the properties of membranes and the activity of enzymes, affects the course of chemical reactions, viscosity of cell sap and the cytoplasm, and of the sap in the vascular bundles. Extreme temperatures also cause changes in the level and activity of hormones. This, on the one hand, is an indication of plant adaptation to the stress arising and a result of injuries, on the other.

11

POLLINATION—FOOD CROPS AND BEES

Ramdane Dris
World Food Ltd.
Meri-Rastilantie 3C
Helsinki, FIN-00980
website: www.food-farming.info or www.world-food.net
e-mail: Dris.uh@Kolumbus.fi

INTRODUCTION

Have your strawberries lost their sweetness? Do your apples taste sour? Did you stop growing cucumbers because they did not produce many fruits and what they did produce were crooked?

There are handy products for the average gardener. Almost all plants have their own fertilizers and all insects have their own pesticides. Every spring, gardeners are encouraged by new seed catalogues and by their supply of seeds controlled for germination. There is one aspect of wise gardening which is yearly addressed. Have you noticed around you that the amount of pollinators vital for plant productivity are decreasing every year?

COOPERATION BETWEEN PLANTS AND POLLINATORS BRINGS BENEFITS

One hundred years ago wild and natural pollinators were abundant. In the 1800s there were plenty of ditches and road shoulders. Natural flora and insects had a lot of space for living. Subsurface drainage systems, modern and extra efficient agriculture have changed the countryside. Extensive fields with just one plant type and the use of pesticides are common these days.

About 80% of the flowering plants are pollinated by insects. In pollination pollen is transported from the anther of the stamen to the stigma of the pistil in the same or another plant. Fertilization is needed for plants to grow seeds.

Pollen is rich in protein. Plants produce nectar when they are most fertile. These two agents together ensure fertilization. Both pollen and nectar are used by insects and their larvae for food. Nectaries are located so that insects feeding on them will be covered with pollen, thus transporting it to other flowers of the same species. This cooperation ensures regeneration for both. Yellow and black wasps enjoy sweet nectar, but they feed their brood with small insects. Wasps destroy a vast number of insects destructive to plants.

LACK OF POLLINATORS IN SPRING

It is difficult to increase the amount of wild and natural pollinators. Their population varies greatly. Only the queen of the wasp and bumblebee colonies hibernates. In early May, the queen starts laying eggs and raising her brood. Due to this there are not many wild and natural pollinators in early summer. A bumblebee or a wasp buzzing in different flowers is usually the queen collecting food for the first larvae.

This is not the case in a honey-bee colony where over 15,000 worker bees spend the winter month with the queen. In spring, the number of bees increases fast. From April the queen lays over 1000 eggs each day. The brood needs fresh nectar and pollen to feed on and flowers close to a bee colony are buzzing with bees from early summer. In midsummer there are already at least 80,000 worker bees in a strong colony. The honey-bee is the most effective pollinating insect. There are many of them from early summer and colonies can be transported to places where plants depend on insect pollination.

IMPORTANCE OF HONEY-BEES IS INCREASING

The importance of bees as pollinating insects has increased all over the world. It is estimated that a quarter of human food depends on bee pollination. There are about 42,000 bee colonies in Finland and this is not enough. At least 4000 more colonies are needed in agriculture alone for seed, fruit and berry production. Honey provides income for the beekeeper but in the national economy it is only a by-product. The value of bee pollination is 6-8 times greater than the monetary value of the honey harvest.

One worker bee makes over ten foraging trips daily and on each trip visits dozens, even hundreds of flowers. During the summer one bee colony makes millions of foraging trips and pollinates billions of seeds. The closer to the plants pollinated by bees the colonies are placed, the greater the crop.

The best distance is less than 500 metres. Bees are loyal to flower species. They seek the best nectar available close by and forage on it. A hectare of clover (*Trifolium*) has tens of millions of flowers that depend on insect pollination. When enough bee colonies are placed close to the area, the amount of seeds increases by 20-30 per cent. A field of turnip rape (*Brassica oleifera*) pollinated by bees produces 10-20 per cent more seeds whose oil content is also superior.

RESEARCHERS ARE INTERESTED IN BEE FORAGING

Studies have been done on the amount and quality of crops pollinated by bees. Areas close to bee colonies are compared to areas depending on other wild pollinators. The yield of crops is compared. Often a third area is included in the study. This area is isolated from insects during the day. This comparison area is thus pollinated by plants themselves or by wind.

Domestic crops close to bee colonies increase an average of 25 per cent and commercial crops 20 per cent. Results in herb cultivation have been encouraging. Herbal subspecies favoured by bees are, for example, caraway (*Carum carvi*), hyssop (*Hyssopus officinalis*) and oregano (*Origanum vulgare*).

There have been similar studies on wild berries and the increase in crops has been remarkable. A study done at Häme Polytechnic College in 1996 investigated the effects of pollination on crops of blueberries and lingonberries (*Vaccinium myrtillus* and *Vaccinium vitis-idaea).* Harvests in the study area tripled for blueberries and lingonberries increased fivefold. In addition, the berries were bigger and tastier.

Beekeepers know these facts through their everyday work with bees. Berries that grow close to beehives are the biggest and most beautiful and they tempt beekeepers on their way back home from beekeeping work. A beekeeper in northern Finland once transported beehives to swamps for two weeks when cloudberries (*Rubus chamaemorus*) were blooming. Bees collect over ten kilos of golden brown honey, with a strong aroma, during this time. Due to bees, cloudberries are so dense they practically form a rug appealing to berry pickers. The cloudberry honey of this area has won the annual award three years in a row for the best honey in Finland.

BEES ARE SOUGHT AFTER FOR POLLINATING WORK

The innovative gardener sprays honey-water on berry blossoms. It is fashionable to plant seeds of traditional plants. Lucerne (*Medicago sativa*), catchfly (*Lychnis viscaria*), harebell (*Campanula rotundifolia*), daisy (*Leucanthemum vulgare*) and cornflower (*Centaurea cyanus*) have re-emerged, pleasing all insects, not to mention beautiful butterflies. Even small areas like this create 'nectar-alleys' which attract insects to the garden.

The best way to improve your crop is to get a couple of beehives. In our northern location, bees are domesticated animals, who need the helping hand of a human to survive the winter. Bees stay in their hive over the 6-7 winter months. They collect tightly in the centre of the hive and by moving their wings create heat which in the centre of this formation can reach +30°C even when on the outside it is -30°C.

Wild nature is plentiful and excellent unutilized fields available in Finland. There is a high demand for Finnish honey, nature's need for pollination increases and new subsidiary sources of livelihood are developing in connection with traditional agriculture. About one-third of our strawberry and black, red and white currant fields are located in eastern Finland. There are about the same number of beehives as there are hectares of berry fields, that is, about 3000. The number of berry growers who can ensure crops with bee pollination is far too few. Good seed production of natural plants also depends on this. Plants produce seeds that are bigger and more fertile. Bees safeguard nature's diversity.

HONEY, A HEALTHY NATURAL PRODUCT MADE BY HONEY-BEES

It is estimated that bees visit 5-7 million flowers in order to produce one kilo of honey. Bees utilize their glands to produce enzymes which convert nectar into honey. The end result is a well-preserved, bacteria-destroying, easily digestible, readily energized food, which has been used by humans for at least 15,000 years. Information on the health benefits of honey is based on research and practical experience over the centuries.

Each flower gives its flavour and colour to honey. The colour of honey ranges from almost white to nearly black. For example, raspberry and clover honeys are almost white and their consistency soft, whereas heather honey is reddish-brown with a jelly-like consistence. Honey collected from different flowers crystallizes according to the type of nectar. Turnip rape honey is yellowish-white and rich in glucose and will crystallize in a couple of days. Willow-herb (*Epilobium angustifolium*) honey is richer in fructose and can stay runny over one year.

There are many other kinds of natural sugars in honey, which are converted by bees into a more digestible form than ordinary cane sugar. Due to this diversity of sugars and aromas, honey is an excellent flavour enhancer in food. It is worth trying flavouring meat dishes or fish baked in the oven with honey.

In addition to these natural sugars, there are micronutrients, minerals and some vitamins in honey. Honey's health benefits are based on the amount of enzymes bees have transferred into honey. Some nectar requires more work from bees to become honey and thus this honey contains more enzymes. The most common honey in Finland is multifloral honey and rich in enzymes. In addition, the best quality honeys based on the amount of enzymes and other characteristics include many one-floral honeys, for example heather, thyme and lime tree. A beekeeper can also specialize in the other natural products obtainable from beehives: pollen, beewax and propolis. Their production and utilization has started well in Finland.

Our self-sufficiency rate in domestic honey in the best of years is only 60-70%, the average national honey harvest is 1.8 million kilos. There is much demand because consumers appreciate Finnish honey, which is produced in an unpolluted environment.

12

EFFECT OF MANURE DOSES AND GROWTH MEDIA ON PRODUCTIVITY AND MINERAL COMPOSITION OF ROCKET LEAVES (*Eruca Sativa*)

Sezai Delibacak[1], Okur Bülent[1] and Dursun Eşiyok[2]
[1]Ege University, Faculty of Agriculture, Department of Soil Science, 35100 Bornova, İzmir, Turkey6
e-mail: delibacak@ziraat.ege.edu.tr3
Fax: +90.232.3881864
[2]Ege University, Faculty of Agriculture, Department of Horticulture, 35100 Bornova, İzmir, Turkey

INTRODUCTION

Rocket (*Eruca sativa*) is a member of the Brassicaceae family. It is endemic in most Mediterranean countries and in Northern and Eastern Europe and thrives in many kinds of soils (preferably calcareous) in fields. Green fresh leaves are mostly used as appetizers with traditional Turkish foods such as 'pide' (Turkish pizza) or 'kīsīr' (Turkish wheat meal), or to prepare fresh salads (Eşiyok, 1996).

Rocket is believed to have aphrodisiac properties (Fernald, 1993). It is used in traditional pharmacopoeia for many different purposes: antiphlogistic, astringent, depurative, diuretic, digestive, tonic, laxative, stomachal discomfort, anti-inflammatory for colitis, antiscorbutic and rubefacient (Bianco, 1994). An infusion at 4-8% is used against itching, chilblain, scalds and urticaria. Among other applications are a lotion to enhance hair regrowth and to counter greasy scalps (Ellison et al., 1980), a tonic for the face (Anonymous, 1988) and a cure for catarrh and hoarseness (Mascagno, 1987). Oil and leaf extracts have also shown good potential as insect repellants (Sharma and Wattal; 1982).

According to 1996 records, the production of rocket is nearly 1100 t per annum in Turkey. It is not possible to assess the exact status of rocket production, because much of it is produced on small parcels and in home gardens, which are generally not included in official records (Eşiyok, 1996).

In this research, the effects of different levels of manure and growth media on the productivity and mineral composition of rocket leaves are considered.

MATERIAL AND METHODS

The experiment was conducted in fields characterized by sandy clay-loam soil (Entisol-Typic Xerofluvent) with slight alkaline reaction. On a volume basis, 20% pumice and 20% perlite were mixed with soil separately to 10 cm depth.

Three different levels of manure (4 ta^{-1}; 8 ta^{-1}; 12 ta^{-1}) were applied to experimental soil alone, experimental soil mixed with pumice and experimental soil mixed with perlite. Five kga^{-1} N; 2 kga^{-1} P_2O_5; 1.5 kga^{-1} K_2O were applied to growth media as basic fertilization at sowing time. The experiment was designed according to randomized blocks with three replications. Thirteen different growth media were used and referred to by numbers: 1-Treated soil + 20% pumice; 2-Treated soil + 20% pumice + 4 ta^{-1} manure; 3-Treated soil + 20% pumice + 8 ta^{-1} manure; 4-Treated soil + 20% pumice + 12 ta^{-1} manure; 5-Treated soil + 20% perlite; 6-Treated soil + 20% perlite 4 ta^{-1} manure; 7-Treated soil + 20% perlite + 8 ta^{-1} manure; 8-Treated soil + 20% perlite + 12 ta^{-1} manure; 9-Treated soil; 10-Treated soil + 4 ta^{-1} manure; 11-Treated soil + 8 ta^{-1} manure; 12-Treated soil + 12 ta^{-1} manure ; 13-Untreated soil (control).

Sowing of the experimental parcels (1 m^2 surface) with seeds of rocket (*Eruca sativa*) was done in quantities equivalent to 2 gm^{-2}. Seeds were sown on 23 September 1997 to 1 cm soil depth, in rows 20 cm apart, without tilling operations. Irrigation was provided when required. First (5 November 1997) and second harvest (10 December 1997) was done manually with knives. The fresh yield of rocket leaves was weighed and the plant material from each harvest than dried in a forced air oven at 70°C for 48 hours. Total N, P, K, Na, Ca (Kacar, 1972), Mg, Fe, Cu, Zn and

Mn (Munoz, 1968) analyses were done of dried rocket leaves. Data were analysed according to the Tarist statistical program (Açīkgöz et al., 1994).

RESULTS AND DISCUSSION

Some physical and chemical properties and mineral composition of the experimental soil are shown in Table 12.1

Table 12.1 Some physical and chemical properties and mineral composition of experimental soil

Sand %	*Silt %*	*Clay %*	*Texture*	*pH*	*$CaCO_3$ %*	*Soluble salt %*	*Org.mat. %*	*Total N %*
52.56	24.00	23.44	SCL	7.42	2.53	0.075	1.93	0.143
P $mgkg^{-1}$	K $mgkg^{-1}$	Ca $mgkg^{-1}$	Mg $mgkg^{-1}$	Na $mgkg^{-1}$	Fe $mgkg^{-1}$	Cu $mgkg^{-1}$	Mn $mgkg^{-1}$	Zn $mgkg^{-1}$
5.44	250	4100	258	40	2.80	5.64	19.98	5.74

The marketable yield and mineral composition of rocket from the 13 different growth media are given in Table 12.2.

As shown in Table 12.2, the highest yield (2717 gm^{-2}) of rocket was found at the first cutting with treated soil + 20% pumice + 4 ta^{-1} manure, followed by treated soil + 20% perlite + 4 ta^{-1} manure (2483 gm^2) and treated soil + 4 ta^{-1} manure (2420 gm^2). The lowest yields of rocket were determined from mixtures of 12 ta^{-1} manure. Thus, increasing levels of manure decreased rocket yields. There was no statistical relationship between second cutting yields of rocket of different growth media. Yield of rocket decreased in the second harvest. Results obtained corroborate those of Bianco and Boart (1996) and Pimpini and Enzo (1996). Total yield of rocket was highest in treated soil + 20% pumice + 4 ta^{-1} manure, followed by other growth media mixed with 4 ta^{-1} manure. Increasing levels of manure decreased total yield of rocket. Comparison of first and second and total yields of rocket in different growth media are shown in Figure 12.1.

It was determined that increasing levels of manure (Table 12.2) increased mineral compositions of rocket samples. Increasing levels of manure caused an accumulation of N, P, Na, Mg, Mn and Zn in rocket leaves. Arguably, decreased rocket yield might be the reason for accumulated mineral composition in rocket leaves.

According to the results, different growth media can be used to produce rocket. It was found that addition of different levels of manure to growth media caused statistically significant variations in respect of

Table 12.2 Marketable yield and mineral composition of rocket from different growth media

Treatments	*Marketable rocket* (gm^{-2})			*N*	*P*	*K*	*Na*	*Ca*	*Mg*	*Fe*	*Cu*	*Mn*	*Zn*
No.	*1st cut.*	*2nd cut.*	*Total yield*	*%*	$mgkg^{-1}$	$mgkg^{-1}$	$mgkg^{-1}$	$mgkg^{-1}$	$mgkg^{-1}$	$mgkg^{-1}$	$mgkg^{-1}$	$mgkg^{-1}$	$mgkg^{-1}$
2	2717	1716	4433	4.444	5097	56202	1472	25854	3081	617	16.6	89.0	47.3
6	2483	1625	4108	4.464	5120	53047	1374	24658	3081	593	15.3	92.3	55.0
10	2420	1730	4150	4.430	5139	53638	1407	24958	3157	543	17.6	94.3	59.0
1	2243	1630	3873	4.153	4825	53836	949	25555	3419	637	13.3	86.6	47.6
5	2117	1435	3551	4.389	4540	52456	982	23463	2630	583	13.6	72.6	41.6
9	1967	1710	3676	4.404	4916	52061	982	24509	3081	579	16.6	88.3	45.3
3	1913	1440	3520	4.487	5255	52850	1538	23613	3119	503	15.0	88.3	49.3
7	1820	1360	2847	4.583	5239	53639	1996	23613	3457	572	15.0	104.0	51.0
11	1617	1785	3402	4.445	5310	54230	2061	25406	3420	523	19.0	106.3	59.0
13	1567	1736	3303	4.201	4916	52258	982	25703	3459	624	18.6	84.3	60.0
4	1550	1370	2920	4.586	5550	56005	2159	24509	3946	517	17.0	101.3	55.0
8	1517	1550	3067	4.629	5628	54033	2159	24808	3833	573	20.3	112.3	50.6
12	1417	1480	2897	4.620	5563	56991	2258	26004	3607	440	18.3	105.0	64.3
LSD													
(0.05):*	—	ns	—	—	—	ns	—	1521.7	—	ns	—	—	—
(0.01)**	671.3	ns	1367.5	0.33	508.9	ns	407.6	—	685.84	ns	3.44	15.63	14.15

ns : not significant

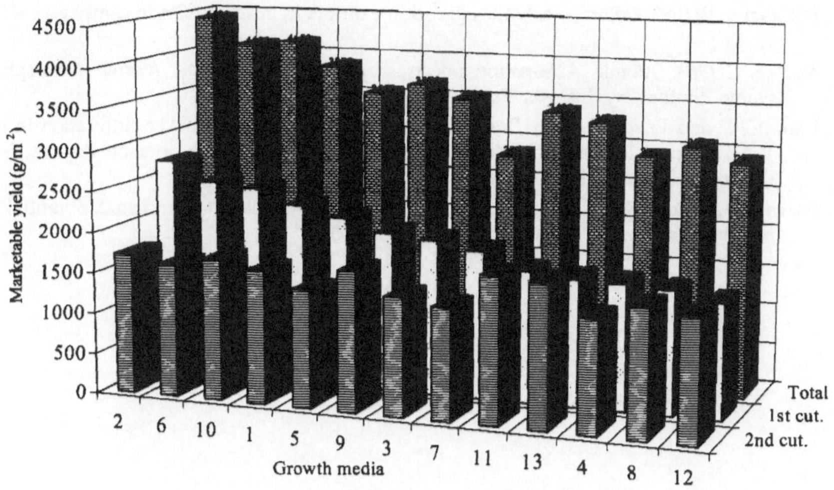

Fig. 12.1 Comparison of first and second cuttings and total yields of rocket in different growth media.

yield and mineral composition of the plant. In particular, rocket yield was negatively affected by the highest levels of manure. Application of 4 ta^{-1} manure could be said to be the optimum level for rocket production.

References

Açīkgöz, N. et al., 1994. TARİST: Tarimsal Araştīrmalar İçin Bir İstatistik Paket Program [A Statistical Packet Program For Agricultural Research] E.Ü.Z.F. Tarla Bitkileri Bölümü, Bornova, İzmir.

Anonymous, 1988. Il grande Libro delle erbe (1st. ed.). Peruzzo Editore, Italy.

Bianco, V.V. 1994. Rocket, an Ancient Underutilized Vegetable Crop and Its Potential. Rocket Genetic Resources Network, Report of First Meeting, 13-15 November 1994, Lisbon, Portugal.

Bianco, V.V. and Boart, F. 1996. Up-to-date developments on world rocket cultivation. Rocket: A Mediterranean Crop for the World. Workshop Report 13-14 December 1996, Legnaro (Padova), Italy.

Ellison, J.A., Hylands, P., A.Peterson, A., Pick, C., Sanecki, K. and Stuart, M. 1980. Enciclopedia delle erbe. (1st. ed.) A.Mondadori, Verona.

Eşiyok, D. 1996. Marketing and utilisation of rocket in Turkey. Rocket: A Mediterranean Crop for the World. Workshop Report 13-14 December 1996, Legnaro (Padova), Italy.

Fernald, M.L. 1993. Gray's Manual of Botany. (1st. ed.) 2. Dioscorides Press, Portland, OR. USA.

Kacar, B. 1972. Bitki ve Toprağīn Kimyasal Analizleri. 2. Bitki Analizleri. Ankara Üni. Ziraat Fak. Yayin no. 453. Ankara.

Mascagno, V. 1987. Coltiva o selvatica la rucola e ottima in insalata. Vita in campagna 5(12): 42-43.

Munoz, J. 1968. Atomic Absorption Spectroscopy and Analysis by Atomic Absorption Flame Photometry. Elsevier Publ. Co., Amsterdam.

Pimpini, F. and Enzo, M. 1996. Present Status and prospects for Rocket cultivation in the Veneto region. Rocket: A Mediterranean Crop for the World. Workshop Report 13-14 December 1996, Legnaro (Padova), Italy.

Sharma, S.K. and Wattal, B.L. 1982. Further studies on mosquito larvicidal potential of mucilaginous seeds. J. Entomol. Res. 6: 159-165.

13

INFLUENCE OF A BIOFUNGICIDE, TRICHODERMIN-BL, ON GROWTH AND DEVELOPMENT OF PLANTS

Dmitry V. Voitka
Belarussian Scientific-Research Institute of Plant Protection,
Minsk region, Belarus 223011

INTRODUCTION

Numerous data are available on the positive influence of soil microorganisms on growth of major plants. Use of preparations with growth-stimulating properties enables more complete realization of the genetic potential of plants.

The fungi of *Trichoderma* (Pers ex Fr.) genus is widely used in agricultural practice for disease control. Apart from the antagonistic properties in relation to disease agents of agricultural crops, a number of works by domestic and foreign authors report that these fungi are able to stimulate plant growth and development.

The addition of dry inoculum of *Trichoderma* spp. in the nutrient substratum accelerated growth and blossoming of annual floral plants. The application of *T. harzianum* accelerated growth of common chrysanthemum roots and shoots, grown from cuttings (MacKenie et al., 1995; Ousley et al., 1994).

Michalikova (1995) noted a decrease in root rot infection, increase in height and dry weight of above ground parts, and number, length and dry weight of hypocotyls of winter wheat roots.

Trichoderma spp. fungi have the ability to induce plant stability, thereby promoting the best growth and development (Harman, 2000).

Trichodermin-BL application in soil during fibre flax sowing produced a yield of 11.6 t ha^{-1} of flax straw (vs control 6.6 t ha^{-1}); and 855 heads per 100 plants (vs control 479). Trichodermin-BL application in the substratum of cucumber seedlings growing in clay pots promoted model plant weight increase in 8 days after the full seedling stage of up to 19.0 g (control 6.5 g). With Trichodermin-BL soil treatment in hotbeds the number of male and female flowers per plant increased to 83 and 14 (vs control 53 and 5 respectively), while yield increase over control was 46.4% (All-Russian, 1965).

Pristchepa and Voitka (1999) and Voitka (2000) have reported the growth-stimulating effect of Trichodermin-BL for tomato and cucumber grown in small-volume hydroponics on substrata of an upland moor.

Products of metabolism of various *Trichoderma* spp. strains are non-identical, attesting to their selective action during the process of agricultural plant growth. The selective character of active products of different *Trichoderma* species and strains is revealed not only within the limits of the plant families which they influence, but also their varieties (Seyketov, 1982).

We evaluated the growth-stimulating effect of Trichodermin-BL (strain-producer of *Trichoderma lignorum* T 13-82) on growth and development of open and protected ground plants.

MATERIAL AND METHODS

Presowing Trichodermin-BL treatment (20 g kg^{-1}) of tomato, beet and carrot seeds was done according to 'Methodical recommendations...' (Mukhin, 1979). The trial variants included presowing seed powdered with the preparation and control (without treatment). The preparation titer was 6 mlrd spores g^{-1}. The power of germination and laboratory seed germination was taken into account.

Spring wheat seeds were treated before sowing. The application rate of a preparation was 5 kg t^{-1} seeds. The trial variants included control (without treatment), Trichodermin-BL (powdering), Trichodermin-BL + greatly swollen polymer hydrogel for adhesion (SSPG).

The biometric parameters were taken into account according to the recommendations of the All-Russian Scientific-Research Institute of Plant Protection (1998).

For cucumber growing in soil and subsoil, Trichodermin-BL was applied as follows: seedlings watered with a suspension of the preparation (2 g/250 ml of water/plant), plants watered twice with the preparation suspension 20 days after the previous treatment (5 g/250 ml of water/plant).

Purple Echinacea plants were treated by Trichodermin-BL 3 times (200 g preparation/10 L water). As a chemical standard azofos, 75% p. (spraying - 100 g preparation/10 L water) was used.

RESULTS

It was found that presowing tomato, sugar beet and carrot seed treatment enhanced their seed qualities (Table 13.1).

Use of the biopreparation for tomato seedling treatment (3 g/100 ml of water/plant) promoted growth and plant development. Plant height in Trichodermin-BL variant was 9.5 cm (vs control 7.1 cm), number of leaves 4.9 (vs control 3.6), and stem diameter in root column area 0.35 cm (vs control 0.27 cm).

Trichodermin-BL application on summer-autumn rotation cucumbers in soil and subsoil showed a yield 11.7% higher than control.

In open ground, Trichodermin-BL was used for presowing spring wheat seed treatment (variety Bango). Application rate of the preparation was 5 kg t^{-1} seeds. Stem length and rootlet root length were recorded.

Results showed that the biometric parameters of wheat plants in variants in which Trichodermin-BL was used and Trichodermin-BL with greatly swollen polymer hydrogel for adhesion were higher than in control (Fig. 13.1).

Evaluation of the stimulating Trichodermin-BL action on purple Echinacea plants showed that the average flower number per plant with trichodermin was 5.8, in the azofos variant 3.6 and in control 4.9.

SUMMARY

The stimulating action of a biological preparation Trichodermin-BL on seed qualities of vegetable crop seed—tomato, fodder beet, carrot—was determined. Trichodermin-BL use in the cultivation of cucumber, tomato and spring wheat produced more active growth and development of plants, promoting their productivity increase. Application of the biopreparation in purple Echinacea plants increased the output of medicinal raw material.

TABLE 13.1 Influence of a biopreparation Trichodermin-BL (*Trichoderma lignorum* T 13-82) on sowing qualities of vegetable crops

Crop	*Trial variant*	*Method of presowing treatment*	*Power of germination, %*	*Laboratory germination, %*
Tomato (variety Kalinka BelSRIVG)	Control (without treatment)	—	72.5	77.5
	Trichodermin-BL (powdering) 20 g kg^{-1} seeds	20 g kg^{-1} seeds + water 0.5 L kg^{-1} seeds	97.5	100.0
Beet (variety Bordeaux)	Control (without treatment)	—	52.5	70.0
	Trichodermin-BL (powdering) 20 g kg^{-1} seeds	20 g kg^{-1} seeds + water 0.5 L kg^{-1} seeds	65.0	77.5
Carrot (variety Kollisto, All-Russian SRIVG)	Control (without treatment)	—	70.0	65.0
	Trichodermin-BL (powdering) 20 g kg^{-1} seeds	20 g kg^{-1} seeds + water 0.5 L kg^{-1} seeds	70.0	80.0

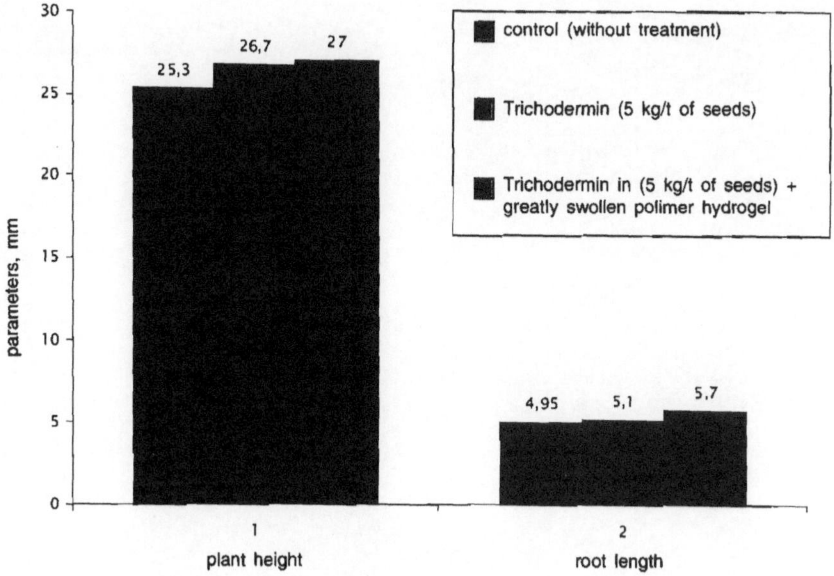

Fig. 13.1 Influence of presowing treatment with a biological preparation Trichodermin-BL on biometric parameters of spring wheat variety Bango plants (field trial, 2000).

References

Alimova, F.K. et al. 1994. Optimization of introduction of a biological safe biopesticide trichodermin in soil against a background of antropogeneous effect, pp. 8-9. In: Introduction of Microorganisms in the Environment. Moscow.

All-Rus. SRIPP, 1965. Application of a fungus trichodermin for disease agents of agricultural crops control. Methodical Manual. Leningrad, 27 pp.

Harman, D.E. 2000. Myth and dogmas of biocontrol. Changes in Perceptions derived from research on *Trichoderma harzianum* T-22 Plant Disease, April, 2000, pp. 377–393.

MacKenie, A.J., Starman, T.W. and Windham M.T. 1995 Enhanced root and shoot of chrysanthemum cuttings infection with the fungus *Trichoderma harzianum*. Hort Sci. 30 (3): 496–498.

Michalikova, A. 1995. The influence of a biopreparation trichonitrin on the growth and development of winter wheat during the seedling stage Acta fytotechn.-Nitra 50: 125–129.

Ousley, M.A., Lynch, J.M. and Whipps, J.M. 1994. The effect of *Trichoderma inocula* on flowering and shoot growth of bedding plants. Sci. Hortic. 59 (2): 147–155.

Pristchepa, L.I. and Voitka, D.V. 1999. Effectiveness of trichodermin in protection of tomato and cucumber against diseases. Bull. Polish Acad. Sci. 47 (2-4): 179–182.

Seyketov, G.Sh. 1982. Trichodermin Genus Fungi and Their Use in Practice. Nauka, Alma-Ata, 248 pp.

Tretiakov, N.N. (ed.). 1998. Physiology and Biochemistry of Agricultural Crops. Kolos, Moscow, 413 pp.

Voitka, D.V. 2000. Possibilities of trichodermin use with cucumber and tomato cultivation by a small-volume hydroponics method. Akhova raslin 5: 30–32.

14

EFFECT OF SOWING SCHEME AND SEED RATE ON YIELD AND QUALITY OF ONION SETS

Zita Barkauskiene
Lithuanian Institute of Horticulture, LT—4335, Babtai, Kaunas distr., Lithuania
e-mail: zita@lsdi.lt

INTRODUCTION

Onions are produced in various ways: from seeds directly drilled in soil, transplanted seedlings or sets (Adamicki, 1998; Rumpel, 2000). Onion sets are onion bulbs under 25 cm in diameter, used to obtain early yield.

In recent years, interest in onion production from sets has been particularly high because such growing can secure the supply of this commodity when onion reserves in stores are exhausted and onions grown from seeds have not yet reached maturity. When growing onions in this way under Lithuanian conditions, the highest, earliest and best quality yield is obtained (Baranauskiene, 1993).

At present, in Lithuania the production of onion sets is insufficient. For a higher yield and quality, the nutritive area and its configuration are of great significance (Brevster, 1994). Different researchers state that in order to obtain the highest output of standard onion sets, taking into consideration seed rate, soil mechanical composition and climatic conditions the seed rate should be 15–25 million ha^{-1} (Brevster, 1994; Rumpel, 2000), because the size of sets depends on the seed rate: the higher the rate, the smaller the sets produced.

Opinions differ regarding sowing schemes. Some authors recommend drilling onion seeds for production of sets in rows with 45 cm between centre points. By this method over 50% of sets bigger than the standard are obtained because plants have too large a nutrient area. Lately, seed belts have been recommended with 15, 20, 25, 27, 30 cm interrows and seeds dispersed in 6–8 cm wide belts (Adamicki, 1997, Rumpel, 2000). This sowing method was not popular, earlier because of the too small assortment of herbicides for onions. Therefore, sowing schemes that facilitated mechanical weed protection were employed. Nowadays, at high herbicide assortment the area under onion sets can be easily protected by chemical means. Thus, at present sowing schemes which provide the highest output of standard sets per unit are of great interest.

This research evaluates the most popular modern sowing schemes of onions for set production and selects the optimal seed rate for the scheme to ensure optimal nutritive area, which preconditions high yield and good quality of onion sets.

MATERIAL AND METHODS

Sowing schemes and seed rates were investigated on the experimental plots of the Lithuanian Institute of Horticulture in 1999-2000. Seeds of the Lithuanian variety 'Lietuvos didieji', germination 90% and 92%, were drilled in the 3rd decade of April.

Trial scheme

A) sowing scheme:

1. (8 – 15) × 4 – 63. In one driving a sowing machine drills 4 seed rows in which seeds are dispersed 8 cm wide. Distances between seed rows 15 cm. Passage for tractor tires 63 cm.
2. (8 – 27) × 3 — 61. In one driving a sowing machine drills 3 seed rows in which seeds are dispersed 8 cm wide. Distances between seed rows 27 cm. Passage for tractor tires 61 cm.

B) seed rates (million seeds per ha): (1) 14, (2) 17, (3) 20, (4) 23, (5) 26.

The trial was carried out in four replications. Record plot length 6 m; width 1.45 m, area 8.4 m^2. Each year after establishing total germination of sets and before harvesting, plant density was recorded in all variants and replications in four places. Biometric measurements were started at the beginning of bulb formation and repeated every ten days until harvesting. For this purpose 20 plants were taken from each variant and replication and the following determined: leaf number and weight, bulb diameter and weight and average values taken.

Sets were harvested when 10% of the onion leaves were lying. Harvested onion sets were dried for 10 days in the field and in the store with active ventilation. Dried sets were cleaned with an onion cleaner, graded into fractions and standard and non-standard onion sets of each fraction calculated. The data obtained was processed by variance analysis.

In 1999, meteorological conditions were favourable. In the spring of 2000 germination was delayed due to drought and before harvesting the precipitation level was higher by 30% compared with multiannual values for this period. Onion sets were grown according to the generally accepted technologies at LIH.

RESULTS AND DISCUSSION

Onion sets grown according to the sowing scheme (8-15) × 4 – 63, gave a higher yield of standard onion sets than sowing scheme (8-27) × 3 – 61 (Table 14.1). According to the first sowing scheme the highest output of standard sets (74.5%) was obtained at the seed rate of 20 million ha^{-1}. The yield of standard onion sets in 1999 was 15.12 t ha^{-1} and in 2000, 14.18 t ha^{-1}. Standard yield slightly decreased (1.5 – 1.76 t ha^{-1}) when the seed rate according to this scheme was 23 million ha^{-1}.

According to the sowing scheme (8 – 27) × 3 – 61, the highest yield of standard onion sets was obtained at the seed rate of 17 million ha^{-1}. In this case the output of standard onion sets was 70.9%.

Quality of onion sets is best characterized by the distribution of standard yield according to size (Table 14.2).

Onion sets of sizes 8–14 mm and 15–20 mm have the highest economic value. The most suitable sowing rates are those which produce the highest output of these fractions. On analysing trial data (Table 14.2) we noticed that the best results were obtained in the first sowing scheme with the seed rate 20 million ha^{-1}. According to the data of two years, the yield of 8–14 mm diameter sets was 5.65 t ha^{-1} and of 15–20 mm diameter 4.62 t ha^{-1}. Insignificantly lower yields were obtained at the seed rate of 23 million ha^{-1}: the yield of 8–14 mm diameter sets was 5.26 t ha^{-1} and of 15-20 mm diameter sets 4.29 t ha^{-1}.

In the second sowing scheme best results were obtained at the seed rate of 17 million ha^{-1}. In 1999-2000 the yield of the smallest fraction was 5.33 t ha^{-1}, and of the medium one 4.26 t ha^{-1}. At the seed rate of 20 million ha^{-1}, the yield of the smallest fraction decreased insignificantly to 5.09 t ha^{-1} and that of 15-20 mm diameter to 4.42 t ha.

Table 14.1 Dependence of onion sets yield on sowing scheme and seed rate (1999-2000)

Seed rate million per ha	*Total yield, t ha^{-1}*	*Standard onion sets yield, t ha^{-1}*	*Output of standard sets, %*
	Sowing scheme (8 – 15) × 4 – 63		
14	17.70	9.83	55.5
17	18.76	11.38	60.6
20	19.71	14.65	74.50
23	19.30	13.02	69.00
26	17.62	10.57	60.00
	Seed scheme (8 – 27) × 3 – 61		
14	16.28	10.50	64.4
17	18.40	13.13	71.4
20	18.02	12.26	68.0
23	18.66	11.42	61.2
26	17.08	9.26	54.2
R_{05}	**1.032**	**0.626**	

Table 14.2 Dependence of yield of various fractions of standard onion sets on sowing scheme and seed rate (1999-2000)

Seed rate million per ha	*Various fractions yield of standard onion sets, t ha^{-1}*		
	8–14 mm diameter	*15–20 mm diameter*	*21–25 mm diameter*
	Sowing scheme (8 – 15) × 4 – 63		
14	1.88	3.39	4.56
17	2.68	4.49	4.26
20	5.65	4.62	4.38
23	5.26	4.29	3.46
26			
	Sowing scheme (8 – 27) × 3 – 61		
14	2.99	3.4	4.1
17	5.33	4.26	3.56
20	5.09	4.42	2.76
23	4.89	4.34	2.14
26	4.64	3.34	1.28

The highest yield of the third fraction with 21–25 mm diameter was obtained at the lowest seed rate of 14 mln ha^{-1}.

The highest yields of standard fractions were obtained in the first sowing scheme with seed rate 20 million ha^{-1} and in the second sowing

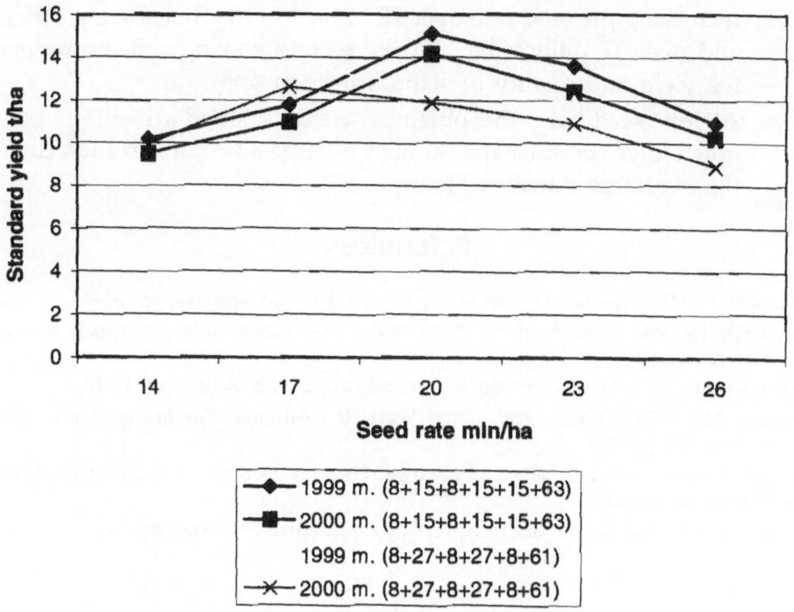

Fig. 14.1 Dependence of onion sets standard yield on sowing scheme and seed rate.

scheme with the rate 17 million ha^{-1} (Fig. 14.1). In addition, these onion sets reached maturity and entered the dormancy period earlier. These factors favourably affected storage capacity and flowering destruction. Further increase in seed rates predetermined a decrease in standard yield and worse maturation. Moreover, bulbs with thick, not fully closed stems were more numerous. Application of smaller seed rates predetermined production of larger-size bulbs, which from the economic point of view are of worse value.

CONCLUSIONS

1. According to the data of two years on production of onion sets of the Lithuanian onion variety *'Lietuvos didieji'* a higher yield of standard sets was obtained according to the scheme (8–15) × 4 – 63 compared to the sowing scheme (8–27) × 3 – 61, because in the first case seeds are dispersed more equally.
2. The highest yield of onion sets was obtained at the seed rate of 20 million ha^{-1} in the first sowing scheme and at the seed rate of 17 million ha^{-1} in the second sowing scheme.

3. Increasing the seed rate over 20 M ha^{-1} in the first sowing scheme and over 17 million ha^{-1} in the second sowing scheme reduced the yield and quality of standard onion sets.
4. At low seed rate, the obtained yield of standard onion sets is much lower because the number of large-size bulbs is high due to the overlarge nutritive area.

References

Adamicki, F. 1997. Przechowywanie warzyw. Wyd. I W. s, Warszawa, pp. 83–91.

Adamicki, F. 1998. Cebula: zbior, dosuszanie, prsechowywanie. Hortpress sp.z.o.o, Warszava, 61 pp.

Baranauskienė, M. ir kt. 1993. Svogūnų ir jų sėklų auginimas. Babtai, pp. 11–14.

Brevster, J.L. 1994. Onions and other Vegetable Alliums. Cambridge Univ. Press, Cambridge, pp. 109.

Rumpel, J. 1998. Cebula–zelecnia uprawowe. Osrodek Doradztwa Rolniezego Olecko, Warszawa, pp. 18.

Rumpel, J. 2000. Uprawa cebuli. Hortpress Sp.z.o.o, Warszawa, 2000. pp. 27.

15

PHYTOPHAGOUS MITES AND THEIR NATURAL ENEMIES IN STRAWBERRY PLANTINGS IN LATVIA

V. Petrova, Z. Čudare, I. Šteinite and J. Tenbergs
Institute of Biology, University of Latvia,
Miera 3, Salaspils LV-2169, Latvia
e-mail: Zigrida@email.lubi.edu.lv

INTRODUCTION

The two-spotted spider mite *Tetranychus urticae* Koch (Tetranychidae) and the strawberry mite *Phytonemus* (*Tarsonemus*) *pallidus* Banks (Tarsonemidae) are the most widespread pests of *Fragaria* × *ananassa* in many strawberry plantings in Latvia (Petrova et al., 2000a). There are many reports of natural enemies feeding on a variety of phytophagous mites. Common predators include lady beetles (Coccinellidae), rove beetles (Staphyliniidae), lacewings (Chrysopidae), true bugs (Anthocoridae), predatory gall midges (Cecidomyiidae), spiders (Phalangiidae, Linyphiidae, Theridiidae), predatory mites (Anystidae, Bdellidae, Phytoseiidae), and some others (Gerson, 1985; Chazeau, 1985; Croft et al., 1998; Tuovinen, 1995).

This study presents information on the abundance, population densities of strawberry pests *T. urticae* and *P. pallidus*, and the identity and abundance of their naturally occurring predators in commercial strawberry plantings in Latvia.

MATERIAL AND METHODS

The study was carried out on field-grown strawberry in the State Horticultural Research Station of Pure located in north-western Latvia. Nearly 50 strawberry varieties are grown there The percentage of infested plants, the population density of phytophagous mites and the fauna of their natural enemies were studied during 1997-2000. A random sample of 100 plants was taken at monthly intervals. The samples consisted of 200 randomly selected (100 folded and 100 unfolded) leaves. Some predatory arthropod individuals, especially adults of anthocorid, coccinellid, chrysopid and spiders were collected by sweep net. One sample comprised insects from 200 sweeps.

RESULTS AND DISCUSSION

Phytophagous Mites

Results of the periodic monitoring of plants infested by *T. urticae* and *T. pallidus* and their population density are presented in Figures 15.1 and 15.2. Usually the percentage of plants infested by *T. urticae* increased progressively to mid-August. As a rule, the change in plant infestation correlated with changes in temperature and the amount of precipitation. In the growing season 1998 with comparatively low temperature, 12.7-17°C, and high amount of precipitation, 29.4-142 mm, the number of infested plants was less than in 1997 and 1999 (Fig. 15.1). Maximum infested plants was observed mid-August 1997 (82.2%). As for population density of *T. urticae,* in 1998 it was moderately and constantly low in number– from 3.3 in May to 2.4 in all stages per leaf in September (Fig. 15.3). Maximum population density was noted in July 1997 (132 in all stages per leaf).

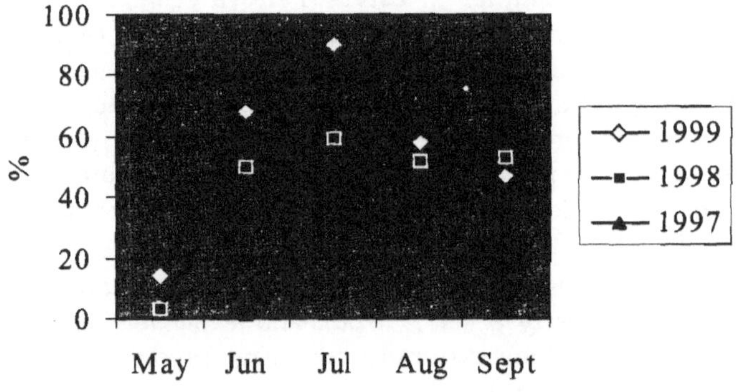

Fig. 15.1 Plant infested by *T. urticae* (%)

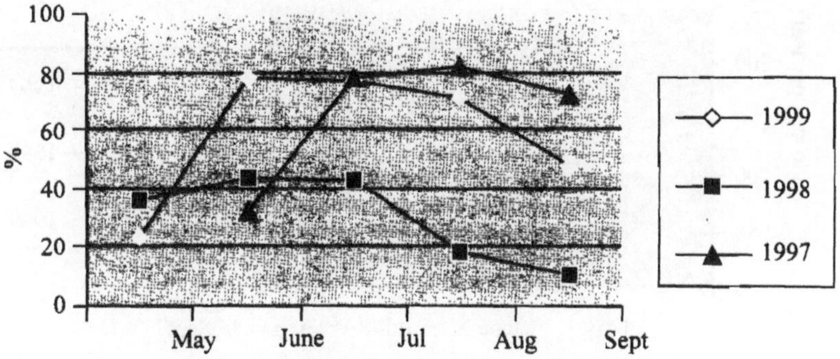

Fig. 15.2 Plant infested by *P. pallidus* (%)

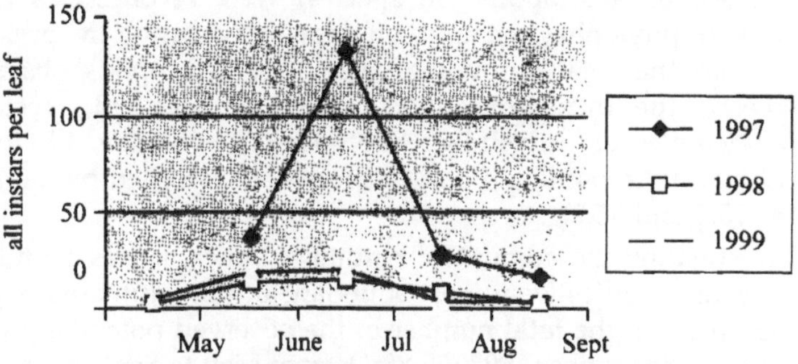

Fig. 15.3 Population density of *T. urticae*

The winter in 1996-1997 was unfavourable for the strawberry mite *P. pallidus*. During the June inspection only 1.2% of all plants were infested, but during the hot subsequent months the percentage of infestation increased to 65.5% in September (Fig. 15.2). The percentage of infested plants was higher in 1999 with maximum in July (90%). Population density of *P. pallidus* was highest in 1997 and 1999, with strawberry mite outbreaks in July and August, up to 211 strawberry mites (all stages per leaf) (Fig. 15.4). In the unfavourable climatic conditions of 1998, *P. pallidus* population density was lower than in 1997 and 1999. During the hottest month, July, only 34.9 pests (all stages per leaf) were recorded.

Natural Enemies

Pest mites can be effectively controlled by naturally occurring enemies on cultivated strawberries in such warmer countries as Spain (García-Marí and González-Zamora, 1999).

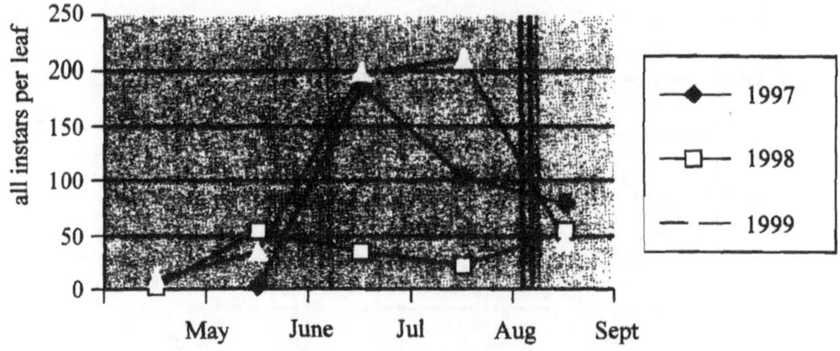

Fig. 15.4 Population density of *P. pallidus*

Predaceous arthropods (48 species) were recorded as potential predators of phytophagous mites on strawberry during the period 1997-2000. From the total amount of beneficial arthropods observed on strawberry, the most investigated were spiders (18 species) and predatory mites of Anystidae, Bdellidae, Cheyletidae, Phytoseiidae, Tydeidae and Aceoseiidae (total 21 species, 17 recorded by the authors) (Tables 15.1 and 15.2).

Investigating the amount of natural enemies, it was found that in total predatory mites were most abundant in June, July and September 1997 (69-100% of the total number of the observed potential predators), from June to September 1998 (55-88%), from May to September 1999 (56-74%), from June to September 2000 (40-64%) respectively. Generally, on the unfolded leaves, where most *P. pallidus* located, phytoseiid (all instars), anystid, bdellid, tydeid mites and larvae of predatory cecidomyiids, and seldom chrysopids and syrphids were found in strawberry mite colonies. On folded strawberry leaves, both pest mites- *T. urticae* as well as *P. pallidus* became targets for other natural enemies, such as anthocorids, coccinellids, predatory thrips and spiders.

The role of phytoseiids as natural enemies of the strawberry mite *P. pallidus* and the spider mite *T. urticae* is important, and their numbers were annually the highest on investigated strawberries. In total, 13 phytoseiid species have been recorded on strawberries in Latvia, i.e., 10 reported by the authors of this paper and 3 (*Euseius finlandicus* Oud., *Amblyseius subsolidus* Begl. and *Phytoseius salicis* Wainst. et Arut.) by Kuznetsov and Petrov (1984).

The assemblage of species and dominant species associated with cultivated strawberry varies in different countries. So, in Russia, 12 phytoseiid mite species have been determined on cultivated strawberry

Table 15.1 List of the predatory mites recorded in strawberry plantings (*recorded in 1997-2000; **recorded by Kuznetsov and Petrov, 1984)

Family	Species
Anystidae	*Anystis baccarum* L.*,**
Cheyletidae	*Cheletomorpha lepidopterorum* Schaw*
Phytoseiidae	*Euseius finlandicus* Oud.**
	Amblyseius agrestis Karg*
	Amblyseius aurescens A.-H.*
	Amblyseius barkeri Hugh.*
	Amblyseius bicaudus Wainst.*
	Amblyseius cucumeris Oud.*
	Amblyseius herbarius Wainst.*
	Amblyseius reductus Wainst.*
	Amblyseius subsolidus Begl.**
	Amblyseius zwölferi Dosse*
	Phytoseius salicis Wainst. et Arut.**
	Proprioseiopsis okanagensis Chant*
	Typhlodromips rademacheri Dosse*
Tydeidae	*Lorryia armaghensis* Baker**
	Lorryia reticulata Oud.**
	Tydeus californicus Banks*
	Tydeus kochi Oud.*
Bdellidae	*Bdellodes longirostris* Herm.*
Aceoseiidae	*Proctolaelaps bickleyi* Bram*

with domination of *A. herbarius* Wainst. and *A. zwölferi* Dosse; in Finland, 15 species with domination of *Anthoseius rhenanus* Oud.; while in Latvia, dominant species are *A. aurescens* A.-H., *A. cucumeris* Oud., and *A. herbarius* Wainst respectively (Tuovinen, 1995; Meshkov, 1996; Petrova et al., 2000 a, b). In Latvian strawberry fields, the anystid mite *Anystis baccarum* L. and phytoseiid mites, *Amblyseius cucumeris*, *A. aurescens* and *A. herbarius* are relatively abundant. Several phytoseiids are potential natural predators of *P. pallidus* or *T. urticae*, but some—*A. cucumeris*, *A. rhenanus*, and *N. fallacis* (Garman)—are potential and specialized predators of both pest mites (Beglyarov et al., 1980; Meshkov, 1996; Zacharda and Hluchy, 1997; Croft et al., 1998). Within the list of predatory phytoseiids found on Latvian strawberry, *A. cucumeris*, *A. reductus* and *A. herbarius* are potential predators of both pest mites, but the most abundant is *A. cucumeris*. Other predatory mites e.g. bdellids, tydeids, cheyletids and aceoseiids, occur on strawberry at low densities. For example, only a few specimens of the cheyletid *Cheletomorpha lepidopterorum* Schaw was found in 1997-98.

Table 15.2 List of spiders and insects—potential predators of phytophagous mites collected on strawberry in 1997-2000

Class, family	*Species*
ARANEA, Theridiidae	*Enoplognatha ovata* Clerck
	Theridon impressum L. Koch
	Theridon varians Hahn
	Theridin bimaculata L.
	Theridium sp.
	Achaeranea sp.
Linyphiidae	*Erigone dentipalpis* Wid. et Reuss.
	Araeoncus humilis Black.
	Bathyphantes gracilis Black
	Oedothorax apicatus Black.
	Stemonyphantes lineatus Clerck
	Silometopus reussi Thorell
	Troxochrus scabriculus Westring
	Tapinocyba insecta L. Koch
	Pelecopsis sp.
	Lepthyphantes sp.
	Walckenaeria sp.
	Neriene sp.
INSECTA, Aelothripidae	*Aeolothrips intermedius* Bagn.
Anthocoridae	*Orius niger* Wolff
	Anthocoris nemorum L.
Chrysopidae	*Chrysopa carnea* Steph.
Coccinellidae	*Coccinella septempunctata* L.
	Adalia decimpunctata L.
	Propylae quatuordecimpunctata L.
Cecidomyiidae	*Arthrocnodax* sp.
	Aphidimyza aphidoletes Rohdani

Altogether nine predatory insect species were identified from strawberry by the authors as potential phytophagous mites. Polyphagous insects such as chrysopids, several coccinellids, *Aelothrips intermedius* Bagn., *Orius niger* Wolff, *Anthocoris nemorum* L. and *Aphydoletes aphidimyza* Rohdani, have been reported as predators of spider mites (Chazeau, 1985; Gerson, 1985). From the observed predatory insects on strawberry plantings the larvae of *Coccinella septempunctata* L., *Adalia decimpunctata* L., *Propylae quatuordecimpunctata* L. (Coleoptera: Coccinellidae), *Chrysopa carnea* Steph. (Neuroptera: Chrysopidae), *Orius niger*, *Anthocoris nemorum* (Heteroptera: Miridae) and *Arthrocnodax* sp. (Diptera: Cecidomyiidae) were found preying on *T.*

urticae. From recorded predatory insects, the gall midge *Arthrocnodax* sp. was associated with populations of both *T. urticae* and *P. pallidus*, therefore it seems to be a specialized predator of both pest mites.

Predatory cecidomyiids and chrysopids were frequent and abundant annually. High numbers of chrysopids were recorded, especially in May 1998-2000 (44-93% from the total amount of beneficial arthropods), and June 1998, 2000 (35-46%). For predaclous cecidomyiids peaks were recorded in August 1997 (58%) and in September 1997, 1998, 2000 (31-35%).

All the other predatory arthropods (spiders, thrips, bugs and coccinellids) were found on strawberry leaves at significantly lower densities and generally in immature stages. Polyphagous spiders of Theridiidae and Linyphiidae can feed on *T. urticae* (Gerson, 1985). From the recorded spiders (18 species) of these families on Latvian strawberry, *Theridon impressum* L. Koch (Theridiidae) and *Silometopus reussi* Thorell (Lyniphiidae) were most common and abundant in samples studied in 1999-2000. However, being less specialized predators of spider mites they were not numerous enough to limit pest mite populations.

In our opinion, phytoseiid mites and other native enemies associated with strawberry are not able to control infestation by both pest mites and cannot keep them at low levels in commercial plantings, especially in warm years. Therefore, only artificial introduction of effective phytoseiids for rapid suppression of pest mites can be an alternative to chemical control in Latvian strawberry plantings.

Acknowledgements

The authors thank Dr. I. Kerzhner of the Zoological Institute, Russian Academy of Sciences, for identification of bugs; Dr. V. Relys of Vilnius University, for identification of spiders; Dr. J. Meshkov of Russian Scientific Research Institute of Phytopathology, for identification of predatory mites. Special thanks to Dr. V. Spungis, University of Latvia, for identification of predatory midges; biologist R. Cinitis, Latvian State Plant Protection Centre, for identification of chrysopids; biologist D. Telnov, of the Institute of Biology University of Latvia, for identification of coccinellids. The study was supported by Latvian Science Foundation grant No. 0284.

References

Beglyarov, G.A., Kondratov, E.S. and Fedotova, N.V. 1980. Effect of prey species on phytoseiid predatory mite development, pp. 37-39. In: N.A. Filipov (ed.) Entomophages of the Plant Pests. Shtiinca, Kishinev. (in Russian)

Chazeau, J. 1985. Predaceous insects, pp. 211-246. In: W. Helle, and M.W. Sabelis, (eds.). Spider mites. Their Biology, Natural Enemies and Control, Vol. 1B. Elsevier, Amsterdam.

Croft, B.A., Pratt, P.D., Koskela, G., and Kaufman, D. 1998. Predation, reproduction, and impact of phytoseiid mites (Acari: Phytoseiidae) on cyclamen mite (Acari: Tarsonemidae) on strawberry. J. Econ. Entomol. 91 (6): 1307-1314.

García-Marí, F. and González-Zamora, J.E. 1999. Biological control of *Tetranychus urticae* (Acari: Tetranychidae) with naturally occurring predators in strawberry plantings in Valencia, Spain. Exp. Appl. Acarol. 23: 487-495.

Gerson, U. 1985. Other predaceous mites and spiders, pp. 375-384. In: W. Helle, and M.W. Sabelis, (eds.). Spider Mites. Their Biology, Natural Enemies and Control, Vol. 1B. Elsevier, Amsterdam.

Kuznetsov, N.N. and Petrov, V.M. 1984. Predatory Mites in the East Baltic Region. Zinatne, Riga. 142 pp. (in Russian)

Meshkov, Y.I. 1996. Phytoseiid mites (Parasitiformes, Phytoseiidae) on main berry plantings. Ph. D. Thesis. Bolshie Vjazemi. (in Russian).

Petrova, V., Čudare, Z. and Šteinite, I. 2000a. Invertebrate fauna on strawberry in Latvia. Proc. Latvian Acad. Sci., Sect. B, 54, 3 (608): 79-84.

Petrova, V., Čudare, Z. and Šteinite, I. 2000b. Seasonal dynamics of predatory mites (Acari: Phytoseiidae) on strawberries in Latvia. J. Ecol. Prob. Biosphere 'Ekologia' Bratislava, 19 (Suppl. 3): 207-210.

Tuovinen, T. 1995. Phytoseiid mites on cultivated berries in Finland, pp. 315-322. In: D. Kropczynska, J. Boczek and A. Tomczyk (eds.). The Acari. Physiological and Ecological Aspects of Acari-Host Relationships. Officyna DABOR, Warszawa, 698 pp.

Zacharda, M. and Hluchy, M. 1997. Biological control of the two-spotted spider mite *Tetranychus urticae* on strawberries by the predatory mite *Typhlodromus pyri* (Acari, Tetranychidae, Phytoseiidae). Exp. Appl. Acarol. 20: 83-94.

16

'FITORODENTS RF'—A NEW ENVIRONMENT-FRIENDLY REPELLENT OF ANIMALS IN GARDENS AND AGROBIOCENOSES

Ivars Zarins[1] and Maris Daugavietis[2]
[1]Latvian University, Institute of Biology,
Miera Street 3, Salaspils, LV 2169, Latvia
[2]Latvian Forest Research Institute
'Silava', Riga Street 111, Salaspils, LV 2169, Latvia
e-mail: maris@silava.lv

INTRODUCTION

Commercial availability of environment-friendly plant protection agents to decrease tree damage by animals is very limited. Tree damage caused by elk, deer, roe hare and small rodents (mice, water-rats or voles etc.) leads to great losses every year.

The search for new effective and cheap repellants is under way. A new environment-friendly repellent based on pine foliage extractives was created and tested in the Institute of Biology and Institute of Forest Research of Latvia. The chemical composition of pine and spruce extractives, physical and chemical properties of 'Fitorodents', taking safety of the environment and soil into consideration, overwintering bark bests, diseases and tree damage by animals, were investigated. Two forms of the repellent—liquid and paste—were tested and the technology for repellent production elaborated.

MATERIAL AND METHODS

Hexane extractives of pine foliage were selected as the more promising for tests. Pine foliage hexane extract contains approximately 50% free acids. Approximately 25% of the extractives are so-called polar acids (diterpene type dicarbon acids and their monomethyl esters), 17% resin acids (diterpene type carbon acids) and 9% fatty acids. Extractives without resin and fatty acids contain chlorophyll and carotene derivates, waxes, natural substances, essential oils etc.

Polar acids and resin acids are considered to be the main animal-repellent substances.

Samples of liquid and paste forms of 'Fitorodents' contain besides extractives 20% (liquid form) or 40% (paste) chalk or kaolin and 1-2% cohesive substances.

The liquid form of 'Fitorodents' is applied on trees by sprinkling and the paste by painting.

The stems and branches of stands of different tree species (apple, pear, cherry, plum) in orchards and birch, ash, oak and maple in afforested agricultural lands were treated with 'Fitorodents' in October-November.

The treated trees ranged in age from 4 to 15 years. 'Fitorodents' was also tested on 1-2-year-old plantings in nurseries.

Trials for protection against burrowing animals during overwintering were also carried out. The liquid form of the repellent was tested on onions, bulbs and other such planting materials for protection against damage by water-rats and moles. Tampons of lignin or flax were placed in the soil around the protected areas.

The effect on plant tissue status after overwintering was evaluated by examining microscopic cuts on bark and buds, and the development of leaves, flowers and root system during vegetation after treatment.

The investigations covered a 5-year period—1997 to 2001.

RESULTS AND DISCUSSION

Investigations revealed the repellent properties of 'Fitorodents' against forest artiodactyls, rodents and other creatures. Since 'Fitorodents' is non-soluble in water, its repellent properties remain effective for 7-8 months. In spring, with enhanced solar heat and air humidity, the colour of the preparation still remains light grey and stems are protected against potential damage due to sharp changes in temperature.

Damages caused to trees by various creatures decreased during the 5-year period by 70-90% (Table 16.1).

Table 16.1 Protective effect of biopreparation 'Fitorodents' (field tests)

Testing	*Pests, creatures*	*Preparation form*	*Protective effect vis-a-vis control, %*
Stem of fruit tree (2-4 year-old plantations)	*Lepus europauns* L.	Solution	92.2 ± 4.75
	Arvicola terrestris L.	"	85.0 ± 4.20
	Apodemus agrarius Pall.	"	72.3 ± 3.35
	Mus musculus L.	"	67.7 ± 3.40
	Apodemus sylvaticus L.	"	82.8 ± 3.80
	Lepus europauns L.	Paste	85.7 ± 4.10
	Arvicola terrestris L.	"	77.0 ± 3.50
	Apodemus agrarius Pall.	"	51.8 ± 2.85
	Mus musculus L.	"	51.80 ± 2.85
	Apodemus sylvaticus L.	"	51.80 ± 2.85
Stem of fruit trees; planting material in nurseries in the wintering period	*Lepus europauns* L.	Solution	80.0 ± 3.30
	Arvicola terrestris L.	"	78.7 ± 3.25
	Apodemus agrarius Pall.	"	69.0 ± 3.00
	Mus musculus L.		
	Apodemus sylvaticus L.		
Propagation material of ornamental plants (bulbs) onions etc. in the substrate	*Arvicola terrestris* L.	Solution	76.3 ± 3.50
	Apodermus agrarius Pall.	"	78.2 ± 4.10
	Mus musculus L.		
	Apodemus sylvaticus L.		
At the storehouse	*Arvicola terrestris* L.	"	70.0 ± 3.30
	Apodemus agrarius Pall.	"	82.8 ± 4.20
	Mus musculus L.		
	Apodemus sylvaticus L.		
Moving passages of creatures,	*Talpa europea* L.	Solution	61.7 ± 3.50
	Arvicola terrestris L.	"	47.5 ± 2.35
	Apodermus agrarius Pall.	"	76.8 ± 3.80
Soil, at depth of 8-18 cm	*Mus musculus* L.		
	Apodemus sylvaticus L.		

Microscopic cuts showed that 1-3% of blossom buds were damaged by the preparation; hence application of 'Fitorodents' to small branches and buds is not recommended. The component in the preparation harmful for buds could not be definitively identified; it might possibly be the solvent (hexane).

Use of 'Fitorodents' decreased the chances of pests and fungal spores to pull through until spring. A better effect was obtained with the paste form of 'Fitorodents' (Table 16.2). Fungicidal properties can be strengthened by adding copper sulphate.

Table 16.2 Protective effect of biopreparation (paste) 'Fitorodents RF' (field tests)

Testing	*Pests, creatures, fungi*	*Preparation form*	*Protective effect vis-a-vis control, %*
Stem and branches of fruit trees	*Lepidosaphes ulmi* L.	Solution	62.5 ± 2.20
	Plesiocoris rugicollis Fall.	"	51.1 ± 2.45
	Lygus pabulinus L.	"	47.3 ± 2.30
	Psylla mali Schmiedl	"	67.0 ± 3.70
	Aphis pomi de Geer	"	56.4 ± 2.50
	Corpocapsa pomonella L.	"	38.0 ± 2.00
	Tmetocera ocellana F.	"	43.4 ± 2.10
	Hyponomeuta malinellus Zell.	"	67.5 ± 2.40
	Panonychus ulmi C.L. Koch.	"	75.8 ± 5.50
	Lepidosaphes ulmi L.	Paste	75.8 ± 5.50
	Plesiocoris rugicollis Fall.		68.3 ± 2.85
	Lygus pabulinus L.		64.2 ± 3.30
	Psylla mali Schmiedl.		78.0 ± 3.45
	Aphis pomi de Geer.		77.5 ± 4.00
	Corpocapsa pomonella L.		57.0 ± 2.80
	Tmetocera ocellana F.		61.2 ± 3.10
	Hyponomeuta malinellus Zell.		67.5 ± 2.40
	Panonychus ulmi C.L. Koch.		63.0 ± 2.75
Splits and cuts on stems of fruit trees	Fungi *Sphaeropsis malorum*	Paste	82.8 ± 3.80
	Fungi *Cryptosporiopsis curvispora*	"	55.7 ± 3.10
	Fungi *Cryptosporiopsis corticola*	"	66.7 ± 3.20

CONCLUSION

1. Extractives of pine foliage soluble in monopolar solvents have good repellent properties against artiodactyls, rodents and other creatures and can be used for production of various kinds of repellents.
2. The preparation 'Fitorodents RF' produced by Biolat company decreases damage to trees by 70-90% (liquid form) or 52-86% (paste).

3. 'Fitorodents RF' has both fungicidal and insecticidal properties and decreases the development of pests overwintering on bark by 38-78%, with better results obtained using paste.
4. 'Fitorodents RF' is effective for protection of onions, bulbs and other such plants against rodents, water-rats and moles. Damage decreased by 48-78%.

17

DIAPORTHE/PHOMOPSIS HELIANTHI, THE SUNFLOWER PATHOGEN: TWENTY YEARS OF RESEARCH

Jelena Vukojević[1], Maria Muñtanola-Cvetković[2] and Sonja Duletić-Laušević[1]
[1]University of Belgrade, Institute of Botany, Faculty of Biology, Takovska 43, 11000 Belgrade, Yugoslavia 6
e-mail: lelabgd@eunet.yu 3 Fax: + 381 11769 903
[2]Institute for Biological Research, Belgrade, Yugoslavia

INTRODUCTION

The first symptoms of a disease of sunflower not known until then, were noticed around mid-July 1980 in the Yugoslav region of Voivodina. The causative agent was described as a new fungal species, *Phomopsis/ Diaporthe helianthi* Munt. Cvet et al. The majority of inbred lines and hybrids were susceptible and the fungus caused severe damage to sunflower plants in Voivodina.

Since 1980, understandably the spread, damages caused by the pathogen, etiology and biology of causative agent, mechanism of pathogenicity, sources of the pathogen, etc. have been studied.

This paper presents a review of this twenty-year period of research.

MATERIAL AND METHODS

Field Material

Sunflower plants were continually collected from the moment of appearance of disease symptoms, during the vegetative period and as detritus. Spread of the pathogen in the host tissues was analysed throughout the vegetative period, after harvest in autumn and in the overwintering period until spring of the following year.

Plant material was collected from numerous localities in Voivodina during these 20 years. Material obtained from France, the Ukraine, Russia and Argentina was also analysed.

Isolations and laboratory studies

Isolations were done from diseased stems and leaf tissues, conidiomata on cankered field-growing sunflower stems, conidiomata on stored seeds, conidiomata on dead stem fragments of sunflower plants and perithecia on overwintered sunflower debris.

Isolation techniques were standard. Growth rates and behaviour of the cultures were followed daily and studied under various light, temperature and nutrient conditions. Standard nutrient media were employed to study the morpho-physiological characteristics of the pathogen, prepared according to Booth (1971): Potato-Dextrose-Agar (PDA), Malt-Agar (MA), Carrot-Agar (CA) and Oat-Agar (OA).

Histological studies: Samples were taken from diseased parts of sunflower leaves, stems and seeds, as well as from stromatic structures formed in artificial cultures, and were embedded in paraffin (Gabe, 1968) for further procedures. Some modifications were necessary because of the nature of the material. Microtome sections 5-8 μm thick were stained with Safranin O and counterstained with Light Green SF Yellowish, which gave the best results in our work.

RESULTS AND DISCUSSION

Appearance and Spread of the Disease

The appearance of a new disease, its unknown symptoms and epiphytotic spread seriously endangered sunflower cultivation in Yugoslavia.

According to reports in the literature, a phytopathogenic fungus belonging to genus *Phomopsis* was isolated from the diseased sunflower stalks in Hungary (Nemet et al., 1981), Romania (Vránceanu et al., 1983), Ukraina (Lesovoy et al., 1993), France (Regnualt, 1985; Lamarque and Perny, 1985), USA – Ohio (Herr et al., 1983), Texas (Yang et al., 1984), Minnesota and North Dakota (Hajdu et al., 1984), Australia (Allen et al., 1980) and Brazil (Homechin and Franca Neto, 1982). The species of genus *Phomopsis* isolated in Voivodina was also isolated from diseased plants in Hungary, Romania, France and Eastern Europe. Morphological and physiological characteristics of numerous isolates from various localities showed no significant differences (Vukojević et al., 1996). *Phomopsis* isolated in the USA (Texas) is morphologically very similar (biotype), while *Phomopsis* isolated in Ohio and Brazil could be associated with already described relatives on soybean and certain weeds (Muntañola-Cvetković et al., 1985a). When the years of appearance of *Phomopsis* in the certain countries are compared, the assertion that the starting point was in the relatively small region confined by the Carpathian mountains in the east and the Tisza and Danube rivers in the west, could be postulated. It is the only region where epiphytotic spreading was noticed. The spread of the disease to the east was slow, with lower intensity.

Twenty years later, breeders of sunflower can select hybrids with high agronomical traits from among the numerous ones tolerant and resistant to *Phomopsis helianthi*.

Origin of the Disease

The appearance of the new sunflower disease caused by an unknown agent led to many assumptions about its origin. Among the cultivated plants, soybean, host to a few species of genus *Phomopsis*, was mentioned frequently.

Mycological study and tests with artificial infections excluded this assumption. The possibility that the pathogen was introduced with seed was also excluded. The amounts of seed are very small, in tens of grams, phytopathologically tested and, furthermore, epiphytotic spreading must be provoked by an enormously high amount of inoculum. Mutations of already existing species were also supposed but this had to be proved by genetic analyses. It could be concluded that the causative agent must have existed in the region on other hosts, and hence weeds were studied as possible hosts (Mihaljčević and Muntañola-Cvetković, 1985).

Compared with the morphological characteristics of the new pathogen, the closest were *Diaporthe arctii* isolated from *Arctium lappa* and *D. arctii* var. *achileae* isolated from *Achillea millefolium*. Tests of cross artificial infections excluded the aforesaid plants as possible inoculum sources (Mihaljčević et al., 1982). In a study comprising 46 weed species from 50 localities in Voivodina, *Phomopsis* was isolated from 15 species but further investigation excluded all of them (Mihaljčević and Muntañola-Cvetković, 1985). No connection between size of weed population, appearance of *Phomopsis* and intensity of symptoms could be established. Further mycological studies denied the assumption that any of the weed species could be alternative hosts to the sunflower pathogen, but confirmed that *Phomopsis* isolated from sunflower is the particular species (Vukojević et al. 1988; Muntañola-Cvetković et al., 1996).

The problem of parasite origin, i.e., the existence of a possible alternative host is still not resolved.

Symptoms of the Disease

In the beginning of July 1980, the first visible symptoms on sunflower plants were noticed. Spots on the stems, at the petiolus base, at first light brown and differing in size, became darker with a wet aspect, increased rapidly in size and after 20 days embraced the whole cylinder. With necrosis of the epidermis, the colour of the spots changed, the dead tissues became lighter and flattened. Dark brown borders of different widths surrounded the affected tissues, on which pycnidial structures could be seen. The inner necrotic process was even more intense and the tissues were soon completely destroyed. With interruption of the water supply and loss of turgor, the plant became moribund. As a consequence of these pathological processes the plants did not attain normal development. The most dramatic effect was a decrease in oil synthesis (Muntañola-Cvetković et al., 1981).

In the following year (1981), the first symptoms were observed on the leaves—necrotic processes arising from the apical end or from margins (Petrov et al., 1981). The fungus spread through the vascular system of the leaf and petiole and entered the stem. The subsequent phase of pathogenesis was characterized by spots on the stems (Fig. 17.1).

The natural infection route was leaf-petiole-stem, as proved at the histological level (Mihaljčević et al., 1985; Bertrand and Tourvieille, 1987; Muntañola-Cvetković et al., 1989).

Causative Agent

When the disease appeared in Yugoslavia, the conidiomata observed at the stem lesions in the field as well as in cultures, containing filiform conidia, led to many wrong assumptions about the identity of the causative agent. Mihaljčević and colleagues (1980) first noticed the conidiomata structures characteristic of genus *Phomopsis*. Although two types of conidia are usually formed by *Phomopsis* species, in the analysed pycnidia only β-conidia were found.

In September 1980, under laboratory conditions during cultivation on PDA perithecia were formed, i.e., the teleomorph—*Diaporthe*. The culture is stored in CMI Kew, England, as exiccata. In spring 1981, numerous ascomata were observed on overwintered stems in the field.

Based on careful investigation of type species and other herbarium material or pure cultures of *Diaporthe* and *Phomopsis* from collections in Baarn, Cornell, Kew and Paris, the new holomorph species named *Phomopsis helianthi* Munt.-Cvet. et al. for the anamorphic stage and *Diaporthe helianthi* Munt.-Cvet. et al. for the teleomorphic phase was described (Muntañola-Cvetković et al., 1981). The parasite belongs to the *D. arctii* group, a large and heterogeneous group of species and forms, with descriptions mostly based on morphological fungal characteristics in the herbarium. Wehmeyer (1933) considered that new varieties and even species could be excluded from this heterogeneous group. Considering that the micromycete isolated from sunflower in Voivodina showed many distinct characteristics, in accordance with the opinion of distinguished specialists, the causative agent was described as a new taxon.

MAIN CHARACTERISTICS OF THE PATHOGEN

The fungus isolated in Yugoslavia in 1980, showed the following characteristics in past research.

Pycnidia

Stem colonization was initiated internally from the mycelium that spread down the petiole from a leaf infection. Hyphae progressed through the vascular system and then spread to the collenchyma and the cortex to the pith. Hyphal masses located between the epidermis and endodermis, where the protopycnidia and pycnidia differentiated. When the pycnidia became mature, egress was achieved when an expanding pycnidium ruptured the epidermis and pushed it aside. In Yugoslav isolates, the pycnidia contained only β-conidia, filling the conidiomata as they became mature, even before the host epidermis ruptured. Pycnidia were also found on sunflower debris (Figs. 17.2-17.4).

Conidia

Abundant and only β-conidia were constantly formed during the whole period of study. Only in 1980 (Muntañola-Cvetković et al., 1980), were α-conidia observed in a certain number of cases, principally on the petiole and rarely on the stem tissue. Aćimović and Štraser (1982) reported abundant α-conidia in cultures of *Phomopsis*, but described a distinct *Phomopsis* species. The existence of a few *Phomopsis* species on sunflower caused confusion in spore presence (Marić et al., 1982; Muntañola-Cvetković et al., 1985a).

It has been proven that germination of β-conidia is a very rare phenomenon and plays no role in spread of the disease. Ultrastructural studies showed that mitochondria were few in number and no reserve polysaccharides or proteins were found in the cytoplasm, which could influence premature senescence and incapacity for germination (Muntañola-Cvetković et al., 1985b).

Perithecia

Perithecia are abundantly formed on detritus in the field. The appearance of ascogonia and development of perithecia are affected by climatic conditions (temperature and humidity) (Vukojević et al. 1991). In field conditions, the nuclear phase change of the fungus from anamorphic to teleomorphic, occurs in the parenchymal cells beneath the endodermis, near the vascular bundles or in the host phloem and cambium. Ascogonia can be observed to form early in autumn. Development of protoperithecia in perithecia and asci and ascospore formation depend on ecological conditions. Under wet conditions or after each rainy period the rostrum of the perithecium elongates and ascospores are released. Ascospores are able to germinate and initiate a new cycle when the growing season of the host plants begins in the field (Muntañola-Cvetković et al., 1989; 1991) (Figs. 17.5-17.7).

Cultural Characteristics

The characteristics of *P. helianthi* cultivated on various nutritive media depended on the medium used. The mycelium on most of the media employed was scantily aerial and white, except on MA where it was more abundant and showed olivaceous shades. Optimal growing temperature was 23-25°C, and rapidity of conidiomata production was mostly 7-10 days. Conidiomata were readily produced, mostly solitary, but sometimes also associated with stromatic tissue, and commonly contained only β-conidia (Muntañola-Cvetković et al., 1985a). Perithecia

Fig. 17.1 Discolorations and cankers surrounding the petiole base.
Fig. 17.2 Fungal hyphae invading the ray parenchyma of the live sunflower stem (Bar = 50 μm).
Fig. 17.3 Hyphal masses (Hm) in the host cortical tissues, as a first step in pycnidial development (Bar = 50 μm)
Fig. 17.4 A mature pycnidium filled with β-conidia erupting through the epidermis of a dead sunflower stem fragment (Bar = 50 μm).
Fig. 17.5 Several ascogonia at various developmental stages (Bar = 50 μm).
Fig. 17.6 Mature perithecia in the ray of parenchyma cells (Bar = 100 μm).
Fig. 17.7 Empty perithecium (Bar = 100 μm).

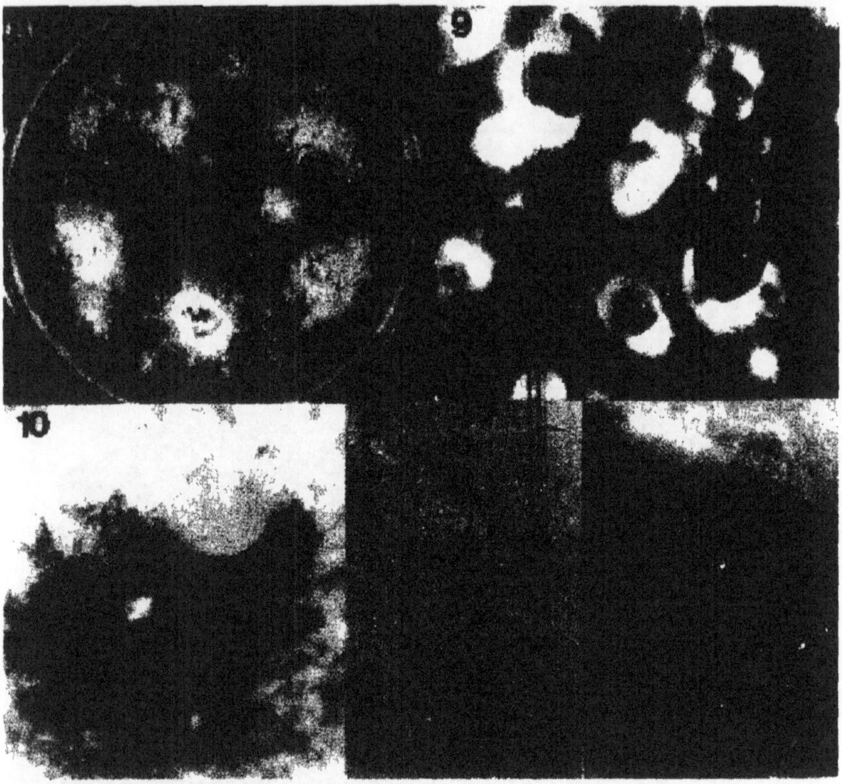

Fig. 17.8 Phomopsis helianthi: 20-day-old colonies on MA.
Fig. 17.9 Simple pycniostromatic structures with drops of exudate (X 20).
Fig. 17.10 Conidiomata associated with stroma, with amber-coloured conidial exudata (X 20).
Fig. 17.11 β-conidia (Bar = 50 μm).
Fig. 17.12 Perithecia on autoclaved stem (X 20).

forming on artificial media were very rare and unstable. The only efficient way to obtain perithecia was usage of the autoclaved stem pieces of various weeds and sunflower and soybean stems (Vukojević et al., 1995) (Figs. 17.8-17.12).

DISEASE CONTROL

Managing disease on sunflower covers a wide range of methods, from trying to avoid or escape it by changing locations or time of sowing, through protection using chemicals, to selection of resistant individuals or lines from heterogeneous populations, to highly developed crossing programs using genes for resistance from cultivated sunflowers or from their wild relatives.

Although some diseases can be controlled by using chemicals, such treatments are usually uneconomical, ecological considerations notwithstanding. Selection of resistant individuals or lines within available populations may serve to improve specific characteristics, but the process is usually slow. To introduce new characters such as resistance to various diseases into adapted cultivars, it is necessary to make crosses with appropriate sources of resistance. Crossing programs use genes for resistance from cultivated sunflowers or from their close annual diploid or more resistant perennial polyploid wild relatives (Seiler, 1992). Sunflower hybrids are exclusively based on a single cytoplasmic male sterility source; hence all hybrids cultivated worldwide are closely related. Although the domesticated sunflower has a narrow genetic basis, sunflower has a large capacity to use wild species in breeding programs, because the genus contains 49 species. The wild species offer a broad genetic base and considerable genetic variability for improvement of the cultivated sunflower, especially for disease resistance. Breeding for resistance is a very effective way to control diseases.

Numerous interspecific hybrids between wild species and cultivated genotypes have been developed in many research centres. Utilization of many species of wild *Helianthus* is limited by low crossability and the high degree of F_1 sterility in interspecific hybrids (Rieseberg et al., 1993).

An important result in breeding for resistance to disease was the discovery of genes for high tolerance to *Phomopsis* using wild species. According to Škorić (1985), several wild annual or perennial species exhibited resistance to the pathogen in field conditions, and the tolerant hybrids were obtained from crosses with cultivated sunflower. Interspecific hybrids between *H. tuberosus, H. rigidus, H. argophyllus* and some others, and cultivated genotypes are rich genetic materials for breeding for resistance to *Phomopsis*. Škorić (1985) found resistance to be partially dominant, while crosses of a highly tolerant line with a susceptible line resulted in a hybrid with an intermediate level of resistance. Genetic studies have shown that sunflower resistance to *Phomopsis* is quantitative, mainly under additive control (Vear et al., 1997). Resistance to *Phomopsis* was associated with the 'stay-green' stem character (Vranceanu et al., 1983) and also correlated with resistance to *Macrophomina phaseoli* and drought (Škorić, 1985).

Classical breeding technology based on sexual crossing requires a long time for introgression of characters. Modern biotechniques can improve the efficiency of breeding. Biotechnology involving modern tissue culture, cell biology and molecular biology offers the opportunity of developing new germplasms, better adapted to changing demands (Friedt, 1992; Alibert et al., 1994). Extensive studies have now been developed on sunflower biotechnology.

CONCLUSION

Research of *Phomopsis/Diaporthe helianthi* proved that plant infection is carried out by ascospores and that the infection route is leaf-petiole-stem. Pycnidia are formed between stem epidermis and endodermis, containing only β-conidia. The nuclear phase change of the fungus from anamorph to teleomorph occurs in the parenchymal cells near vascular bundles, or in the phloem and cambium. In Yugoslavia, perithecia appear abundantly on overwintered sunflower stem debris. Climatic conditions influence the development of perithecia. During the past period numerous isolates obtained from sunflower attacked by *P./D. helianthi* showed constant morphophysiological characteristics on various nutritional media.

Increased breeding efforts are employed in order to broaden the genetic variability of cultivated sunflower because it lacks genes for resistance to most sunflower diseases. Successful applications of new methods of biotechnology can contribute to overcoming the problems of a narrow genetic base and concomitantly accelerate the breeding process.

References

Aćimović, M. and Štraser, N. 1982. *Phomopsis* sp.—novi parazit suncokreta (*Phomopsis* sp.—A new parasite in sunflower). Plant Protection 33: 117-158.

Alibert, G., Aslane-Chanabe, C. and Burrus, M. 1994. Sunflower tissue and cell cultures and their use in biotechnology. Plant Physiol. Biochem. 32: 31-44.

Allen, J. S., Brown, F. J. and Kochman, K. J. 1980. The incidence of sunflower disease in New South Wales and Queensland during the 1978-1979. season. Proc. IV Australian Sunflower Workshop, pp. 1-7.

Bertrand, F. and Tourvieille, D. 1987. *Phomopsis* de tournesol: test de selection. Inf. Tech. CETIOM 98: 12-18.

Booth, C. 1971. Fungal culture media, pp. 449-494. In: C. Booth (ed.). Methods in Microbiology, vol. 4. Academic Press, London.

Friedt, W. 1992. Biotechnology in sunflower breeding–Methods, applicability and prospects. Proc. 13th Int. Sunfl. Conf., Pisa, Italy, pp. 1383-1396.

Gabe, M. 1968. Techniques histologiques. Messon et Cie, Paris.

Hajdu, F., Baumer, J.S. and Gulya, T. 1984. Occurrence of *Phomopsis* stem cancer in Minnesota and North Dakota. Proc. Sunflower Research Workshop, Bismarck, North Dakota, p. 15.

Herr, L.J., Lipps, P.E. and Watters, B.L. 1983. *Diaporthe* stem canker of sunflower. Plant Disease 67: 911-913.

Homechin, M. and Franca Neto, J.B. 1982. *Phomopsis* sp. um novo patógeno do girassol. Empresa Brasileira de Pesquisa Agropecuaria, Londrina, PR. Resultados de pesquisa de girassol, p. 665.

Lamarque, C. and Perny, R.A. 1985. Nouvelle maladie du tournesol. Le Phomopsis. Cultivar 179: 57-59.

Lesovoy, A.I., Parfenyuk, A.I. and Vasilyuk, L.A. 1993. *Phomopsis* of sunflower in the Ukraine and the development of breeding methods for resistance. Arch. Phytopath. Pflanz. 28: 321-328.

Marić, A., Maširević, S. and Shu Lee .1982. Prilog proučavanju *Phomopsis* spp. (*Diaporthe* sp.) prouzrokovača sive pegavosti stabla suncoreta. Plant Protection 33: 403-419.

Mihaljčević, M. and Muntañola-Cvetković, M. 1985. Responses of sunflower plants to different *Phomopsis* isolates from Sunflower. Proc. XI Inter. Sunfl. Conf., Mar del Plata, Argentina, pp. 413-418.

Mihaljčević, M., Petrov, M. and Muntañola-Cvetković, M. 1980. *Phomopsis* sp., a new parasite of sunflower in Yugoslavia. Savr. poljoprivreda 28: 531-539.

Mihaljčević, M., Muntañola-Cvetković, M. and Petrov, M. 1982. Further studies on the sunflower disease by *Diaporthe* (*Phomopsis*) *helianthi* and possibilities of breeding for resistance. Proc. X Inter. Sunfl. Conf., Sydney, Australia, pp. 157-159.

Mihaljčević, M., Muntañola-Cvetković, M., Vukojević, J. and Petrov, M. 1985. Source of infection of sunflower plants by *Diaporthe helianthi* in Yugoslavia. Phytopath. Z. 113: 334-342.

Muntañola-Cvetković, M., Mihaljčević, M. and Petrov, M. 1981. On the identity of the causative agent of a serious Phomopsis-Diaporthe disease in sunflower plants. Nova Hedwigia 34: 417-435.

Muntañola-Cvetković, M., Mihaljčević, M., and Petrov, M. 1985a. Comparisons of *Phomopsis* isolates obtained from sunflower plants and debris in Yugoslavia. Trans. Br. Mycol. Soc. 85(3): 477-483.

Muntañola-Cvetković, M., Bojović-Cvetić, D. and Vukojević, J. 1985b. An ultrastructural study of α-and β-conidia in the fungal genus *Phomopsis*. Cryptogamie, Mycol. 6: 171-184.

Muntañola-Cvetković, M., Vukojević, J. and Mihaljčević, M. 1989. Pathohistology of sunflower stems attacked by *Diaporthe helianthi*. Can. J. Bot. 67: 1119-1125.

Muntañola-Cvetković, M., Vukojević, J. and Mihaljčević, M. 1991. The systemic of the sunflower disease caused by *Diaporthe helianthi*. Can. J. Bot. 69: 1552-1556.

Muntañola-Cvetković, M., Vukojević, J. and Mihaljčević, M. 1996. Cultural growth patterns and incompatibility reactions in *Diaporthe* and *Phomopsis* populations. J. Phytopathology 144: 285-295.

Muntañola-Cvetković, M., Petrov, M., Vukojević, J. and Mihaljčević, M. 1980. Further studies on the Phomopsis-Diaporthe disease of sunflower. Arch. Biol. Sci. 32: 5-6.

Nemet, F., Princzinger, G. and Vörös, J. 1981. New diseases of sunflower in Hungary. Magyar Mezogazdasag 48: 19-11. in Hungarian.

Petrov, M., Muntañola-Cvetković, M. and Mihaljčević, M. 1981. New findings on the disease of sunflower plants caused by *Phomopsis helianthi* Munt.-Cvet. et al. (*Diaporthe helianthi* Munt.-Cvet. et al.). Arh. Biol. Sci. 33: 13-19.

Regnualt, Y.1985. Premieres observations sur le phomopsis du tournesol. Bull. Cetiom 92: 13-20.

Rieseberg, L., Choi, H., Chan, R. and Spore, C. 1993: Genomic map of a diploid hybrid species. Heredity 70: 285-293.

Seiler, G. 1992. Utilization of wild sunflower species for the improvement of cultivated sunflower. Field Crop Res. 30: 191-194.

Škorić, D. 1985. Sunflower breeding for resistance to *Diaporthe-Phomopsis helianthi* Munt. Cvet. et al. Helia 8: 21-24.

Vear, F., Garreyn, M. and Tourvieille, D. 1997. Inheritance of resistance to phomopsis (*Diaporthe helianthi*) in sunflower. Plant Breeding 116: 277-281.

Vránceanu, A.V., Csép, N., Pirvu, N. and Stoenescu, F.M. 1983. Genetic variability of sunflower reaction to the attack of *Phomopsis helianthi* Munt.-Cvet et al. Helia 6: 23-25.

Vukojević, J., Franić-Mihailović, D. and Mihaljčević, M. 1991. Verification of the development of the *Diaporthe helianthi* teleomorph during winter 1989-1990. Plant Prot. 42: 109-117.

Vukojević, J., Franić-Mihailović, D. and Mihaljčević, M. 1995. In vitro production of perithecia of *Diaporthe helianthi*. Mycopathologia 132: 21-25.

Vukojević, J., Franić-Mihailović, D. and Mihaljčević, M. 1996. The comparative investigation of *Phomopsis helianthi* isolates from different European localities. !4th Inter. Sunfl. Conf., Beijing, China, pp. 734-738 (Plant Pathology).

Vukojević, J., Mihaljčević, M. and Muntañola-Cvetković, M. 1988. Cluster analysis of morphological data in the fungal genera *Diaporthe* and *Phomopsis*. Proc. XII Inter. Sunfl. Conf., Novi Sad, Yugoslavia, II, pp. 106-107.

Wehmeyer, L.E. 1933. The genus *Diaporthe* Nitschke and its segregates. Univ. Mich. Stud. 9: 349.

Yang, S.M., Berry, R.W., Luttrell, E.S. and Vongkaysone, T. 1984. A new sunflower disease in Texas caused by *Diaporthe helianthi*. Plant Disease 68: 254-255.

18

ADVANCED TECHNOLOGY FOR FIXATION AND CULTIVATION OF SANDY SOILS AND OTHER UNPRODUCTIVE LANDS

Eftimie Nitescu[1], Vasile Cracium[1], Stefan Popescu[1], Iuliana Lupu[1] and Ilie Filipescu[2]
[1]Gh. Asachi" Technical University of Iasi, Romania
2Agromec "Movileni" S.A., Iasi County, Romania
e-mail: enitescu@hidro.tuiasi.ro; veraciun@ma.tuiasi.ro

INTRODUCTION

Sandy soil fixation and cultivation are two major measures in the fight against the desertification phenomenon in arid zones (Africa, Asia, Australia and North America). These measures must also lead to the inclusion of sandy, less productive or unproductive soils in the agriculturally productive circuit, which has to feed a continuously growing population. Major investments might and should be made in sandy soils located in the environs of large cities.

In southern Romania there are over than 400,000 hectares of sandy, less productive or unproductive soils situated in areas with a high population density (on average, there are 0.5 hectares of agricultural soil per inhabitant). These soils are located in the dry areas of Romania where the annual precipitation average is 300-500 mm while potential evapotranspiration goes beyond 700 mm.

Therefore, our research team worked out a cultivation technology that would create an optimum microclimate favourable to plant growth, in order to compensate the deficiencies of such sandy, less productive soils, namely:

— intensive wind and hydric erosion;
— low water retention capacity (6-10%);
— high or very high water permeability;
— rapid runoff of fertilizers and nutrients needed for plant growth, a process which determines the pollution of phreatic waters;
— nutrient and microelement deficiencies favoured by the low content in organic material (0.2-0.5%).

MATERIAL AND METHODS

Hydro-ameliorative Support

A hydro-ameliorative support (Fig. 18.1) was manufactured in the shape of a composite band with a layered structure consisting of the following:

- A base layer that offers mechanical strength to the band necessary for its mechanized introduction into the soil. After many trials, a biodegradable and hydrophilic non-woven material was selected as the base layer, with a high porosity that allows easy penetration by plant roots.
- A superabsorbent layer of polymeric hydrogel type, which absorbs the water from irrigation or rainfalls. Hence, the water will be supplied in small amounts, only those necessary to hydrogel formation, thus reducing water loss at depth. Depending on need, water will be provided directly from the hydrogel to the root system. Also, the non-woven material incorporated in the band structure can provide a reserve amount of water. A mixture that contains nitrogen, phosphorus and potassium salts as well as growth-stimulating agents of mineral and organic nature were included in the hydroabsorbent of the Terra-Cottem type (Belgium make). If some hydroabsorbents with no supplements are used, the fertilizers and growth stimulants can be applied by a dry-spraying process on the surface of the base layer.
- The seed substratum of a band shape (of 10-15 mm width) is made of porous paper with one face covered with an ecological adhesive. The installation described in Figure 18.1 has been used for seed fixation on the surface of the paper band.

  The feed system lets the seeds fall and then a rubber roll ensures their adherence to the adhesive surface of the band (2).

The period 'b' between seed falling, depending on the agrotechnical needs of plants, is adjusted by the position sensor (5) that commands the distributor (4). The seed band wrapped by the installation on the roll (7) is unwrapped and glued on the surface of the base layer. The position of the seed supporting paper band depends on the height of the upper support part in the air, used as a screen against erosion, and on the plant-specific sowing depth (see Fig. 18.1).

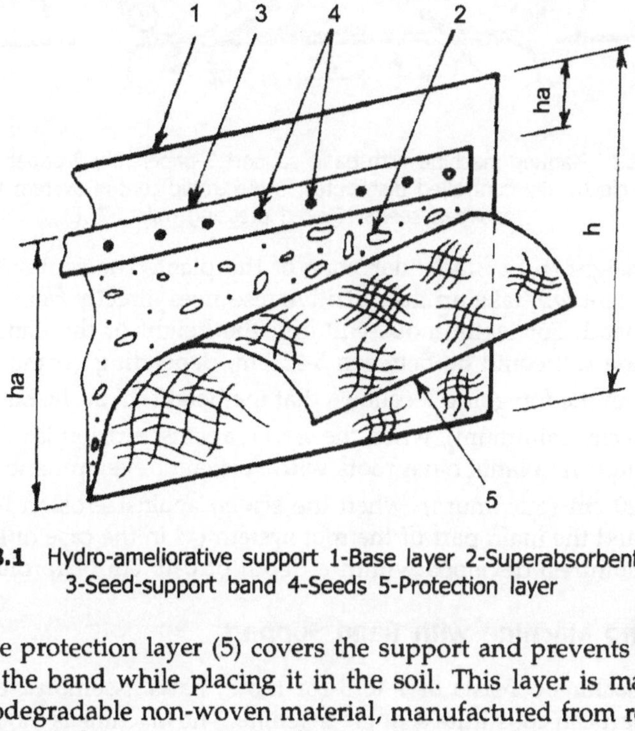

Fig. 18.1 Hydro-ameliorative support 1-Base layer 2-Superabsorbent layer 3-Seed-support band 4-Seeds 5-Protection layer

- The protection layer (5) covers the support and prevents damage to the band while placing it in the soil. This layer is made of a biodegradable non-woven material, manufactured from re-usable fibres at low costs. A needle punching machine was used to interweave the protection and base layers in order to keep the optimum distribution of hydro-ameliorative support elements.

The hydro-ameliorative support (Nitescu, 1999), performed with this technology, is a precast product in the shape of rolls (see Fig. 18.1).

The width (h) of the band support depends on:

— necessary height (h_a) of the air part used as a screen against erosion which might be maximally 50 mm thick to preclude bending (the screen will not be necessary for fixed sands and small slope soils); and

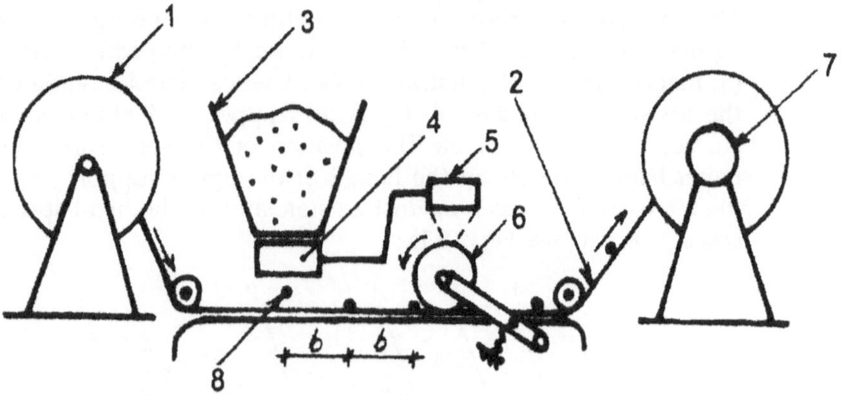

Fig. 18.2 Planting machine with band support 1-paper roll; 2-paper band; 3-seed tank; 4-electrically controlled distributor; 5-electronic control system for interval between seeds; 6-seed pressing roller; 7, 8.

— height (h_s) of radicular zone of the plant. The root system of the plant will take up the nutritive resources directly from the support band. For leguminous cultures, the height of the band buried in the soil could be between 5-15 cm, depending on the plants.

From the foregoing it follows that the height 'h' of the band must be:

— 5 cm (minimum), when the screen against erosion is not necessary and the plants have roots with surficial development;

— 20 cm (maximum), when the screen against erosion is necessary and the main part of the root system (as in the case of leguminous cultures) develops within a depth of 0-15 cm (Caproiu, 1992).

Planting Machine with Band Support

This machine consists of 1 to 3 (or more) parts assembled on a frame. Each part can constitute a single machine. The machine is mounted at the tractor rear. The manner in which the individual equipment is connected with the frame or with a common crossbeam enables alteration of the intervals between the individual equipment and thus of row distance, a function of the vegetable planted. Considering an individual equipment, which represents a single-row machine, this machine has a structure similar to automatic transplanting machines (Bernacki, 1972).

The single-row machine (Figure 18.2) consists of the frame (1) on which are fitted the mounting system (2), adjustable ground wheels (3), press wheels (4) and the furrow opener (5). On the same frame are fitted a support (7) for the band roll (8) and a system which changes the movement of the band roll (8) in a vertical-longitudinal plane from the vertical direction to a horizontal one. This system consists of the inclined guiding roll (6) (fitted in the furrow opener) and the guiding slit (9).

The furrow opener (5) forms a furrow with a cross-section approximating a rectangle, the band-support (8) enters the furrow and the two press wheels (4) intromit it in the soil in a vertical-longitudinal position.

This machine is mounted at the tractor rear and has a working speed of 4-6 km h^{-1}.

PREVIOUS EXPERIMENTAL RESULTS AND DISCUSSION

The composite materials of the support band used for arid or unproductive land cultivation such as sandy soils were studied and tested in the Irrigation Laboratory and Non-woven Materials unit pilot. The technology of mechanized introduction in soil and the machine prototype were executed at the Agricultural Machinery Laboratory.

The base and protective layers are made of hydrophilic non-woven materials manufactured from reusable, organic and biodegradable fibres.

The band has a breaking strength in the longitudinal direction of 11.16 daN and an average elongation of 30%, which allows mechanized introduction in the soil. The hydro-ameliorative material of the band is Terra-Cottem, a mixture of granules with a maximum diameter of 4 mm uniformly glued with an adhesive on the support layer.

The plant seeds for cultivation are inserted in the support band structure. Grass and grain seeds fixed on a narrow paper band (of 15 mm width) were used for experiments with good results, employing the installation prototype presented above. The cultivation machine with support band was designed and performed at the level of prototype. Final construction and experimental results will be published at the end of 2002.

SUMMARY

1. The research team proposed a new fixation and cultivation technology of unproductive lands (sandy soils).
2. The technology includes the manufacturing of a hydro-ameliorative support that creates optimum microclimate favourable to plant growth. The plants germinate and grow inside the band having, at the same time, the possibility by the radicular

system to penetrate the band and grow outside. The support band is introduced into soil in a vertical-longitudinal plane, the upper part of the band remaining outside the soil and playing the role of a screen against wind and hydroerosion.

3. The whole technological process, starting with manufacturing of the support band and continuing with its intromission in the soil was automated and mechanized, and assures increased productivity (over 0.5 ha h^{-1} for cultivation on the band machine).

References

Bernacki, H. 1972. Agricultural Machines, Theory and Construction, vol. I. NTIS, U.S. Dept. of Commerce, Springfield, VA. USA.

Caproiu, S. 1992. Soil Cultivation Machines. Didactic and Pedagogic Publ. House, Bucharest, pp. 109-115. (in Romania).

Nitescu, E. 1999. Hydro-Ameliorative Support. Patent awarded the Gold Medal at The Invention Saloon "EUREKA ", Brussels, Belgium.

19

ANALYSIS OF DEVICES FOR PLANTING SEEDLINGS

Ovidiu Balan
Universitatea Tehnică "Gh.Asachi" Iaşi—Facultatea de Mecanică, Bdul. Mangeron nr.63-65 Loc.Iaşi, cod 6600, Romania
e-mail: echichi@ma.tuiasi.ro

INTRODUCTION

Growing crops by means of manpower systems is particularly practised for vegetable crops such as tomato, green pepper, egg-plant, cucumber, cabbage, celery, onion, leek, cauliflower, turnip, lettuce and potato, but is not uncommon for some technical plants (tobacco, sugar-beet and carrot for seeds), some species of flowers, fruit trees and grapes.

The planting procedure for any crop must consider economic efficiency, especially for vegetable crops, and hence manual labour versus mechanization of this activity given careful thought.

Taking the need for manpower in growing cereals as a comparative base, the situation of manpower expenditure is presented as follows: cereal plantation 100%; fruit orchards 150%; vineyards 305%; vegetable gardens 378%.

It can be seen that use of manpower is already very high. Nonetheless mechanization of vegetable gardening is decreasing in many countries and almost all vegetable growing done manually.

Producing and planting seedlings is very profitable but involves high manpower expenditure and supplementary expenses. For greater profitability mechanization of planting work is essential.

By mechanizing planting work, manual labour can be reduced three or four times, productivity increased because plantation can be begun in the most favourable period, uniform distribution of plants in rows is achieved, and the rows are straight and equidistant enabling mechanized maintenance of crops.

The planting process consists of regular distribution of seedlings–one by one–on the surface unit, intromission and fixation in the soil. Generally, mechanized planting of seedlings takes into account opening of the furrow, individual distribution of seedling drains, vertical fixation of the seedling in the soil and, if necessary, its watering.

All these steps are done by specific working tools, with which all sectors of transplanting machines are equipped, Thus, opening of the drain is done by a ploughshare, intromission of the seedling by the planting device seedling fixation in the soil by pressing wheels and watering by the watering device (Fig. 19.1).

Fig. 19.1 Technological scheme of the transplanting machine.

By moving the machine, the ploughshares (1), having a distance between them equal to the established distance between rows, enter the soil and open furrows at the necessary depths for planting the seedling root-end down. The planting device (2), usually activated by a wheel of the machine, periodically transports a seedling to the supply place and plants it vertically with the root in the furrow opened by the ploughshare. The frequency with which the seedlings are directed into the furrows and the speed of the machine influences the distance between plants in the row. After the ploughshare has passed, the walls of the canal fall in, covering the seedling root. Pressing wheels (3) simultaneously ensure complete and uniform covering of the roots and by compacting the soil around them complete fixation of the plant in the soil.

Thus in a modular planting machine the component parts of one or more planting sections mainly include a ploughshare, pressing wheels and a watering device.

The main unit of a planting machine is the planting device whose functions consist of:

— supplying seedlings to devices called receivers which fix the seedlings in the supplying position;

— moving the seedling from the supplying system to the groove opened by the ploughshare;

— releasing the seedling vertically root downward.

Material and Methods

The planting device is the main element of any planting machine. From the construction viewpoint, the planting devices comprise:

— discs which may have receivers affixed to them and be of the double-disc type;

— receivers placed on different types of conveyors (chain, belt, shelf elevators with pneumatic or gravitational scraping devices);

— special receivers of the multidigital or needle type (rotating needles with extracting bar, needles fixed on the pushing bar or tilting needles).

According to the type of adjustment of distance between plants in the row, planting devices can be classified as follows:

— by changing the number of receivers on the device;

— by adjusting the transmission proportion of the movement of the activating mechanism;

— by modifying the length between the receivers fixed on different conveyors;

— by following an acoustic or visual signal.

Activation of the planting devices is usually done by means of cogwheels, chain wheels or belt wheels centralized from the transportation wheels and individually from the pressing wheels.

The main element of a planting device is a conveyor, which has receivers attached for seedling fixation. The most used receivers are clamp-shaped for simple seedlings and cup-shaped for seedlings grown in nutritive pots.

The most used planting devices are those with receivers fixed on the disc (Fig. 19.2) (Ciubarin, 1972; Balan, 1998a, b).

These devices achieve a uniform distribution of seedlings in the row due to the equal distance between receivers. Also, the quality of the planting system is less influenced by wheel sliding or by adjustment of transmission because the distance covered by the furrow is very short.

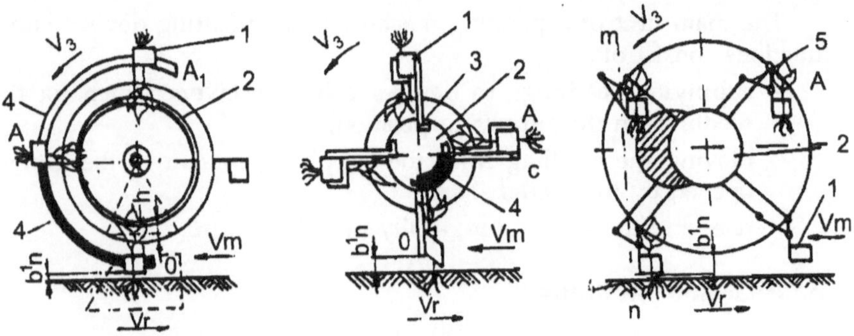

Fig. 19.2 Planting devices with receivers fixed on disc.

Receivers (1) may be clamp-shaped with sponge or rubber protection inside (Fig. 19.2c) cup-shaped.

Receivers may be of the normal open type (Fig. 19.2a) or normal closed type (Fig. 19.2b, c).

Receivers are placed radially on the disc (2) (Fig. 19.2a, b) or by means of levers (5), which keep them permanently in a vertical position.

The clamps which form the receiver (2) are attached on the disc or between discs by means of springs similar to hinges. The springs keep the clamps apart in the case of normally open receivers and close in the case of normally closed ones.

Between the moment of seedling supply and the moment of seedling release in the furrow, the receivers must be opened. Hence the normally closed type have levers with roll (3) which follow the profile of a cam (4) which controls this operation. Insertion of the seedling in the receiver is done horizontally in point (A) or vertically in point (A) or vertically in point (A_1).

Fixation of the seedling between the disc and the clamp, with normal open receivers, is done by means of the cam (4) on which the mobile clamp steps.

The disc is activated in a rotation movement by a copy wheel or transport wheel by means of transmission. The seedling (together with the receiver) moves at a 180^0 or 90^0 after a cycle obtained by aligning the rotation movement of the disc with the forward rectilinear movement of the machine.

The lower part of the receiver (1) opens, releasing the seedling into the furrow cut by the ploughshare. For good planting, the seedling must be intromitted vertically into the soil. It can be seen that in planting devices with radial receivers (Fig. 19.2a, b) the seedling is placed vertically in a very short period of time. This is both a disadvantage—the seedling must be released only during this interval–and also an advantage when the peripheral speed of the receiver having a seedling is not equal to the speed of the machine (in opposite directions). Nonetheless the planting process is possible without damaging the seedling.

The route of the roots into the soil of seedlings emerging from receivers fixed in a permanent vertical position, is closer to a straight line; hence seedling release is done in better conditions, but the speed of the machine must be as close as possible to the speed of the root of the seedling in order to avoid rubbing against the walls of the furrow.

The opening of clamp-type receivers can be done in a direction opposite to the disc rotation (Fig. 19.2.a) or in the direction of disc rotation (Fig. 19.2b).

To plant seedlings grown in nutritive plant-pots, cup-shaped devices fixed on discs are frequently used (Fig. 19.3b) or radial arms (Fig. 19.3a) which rotate horizontally or vertically. It is important that during movement of the seedling, the cups be kept in a vertical position, perpendicular to the forward movement of the machine.

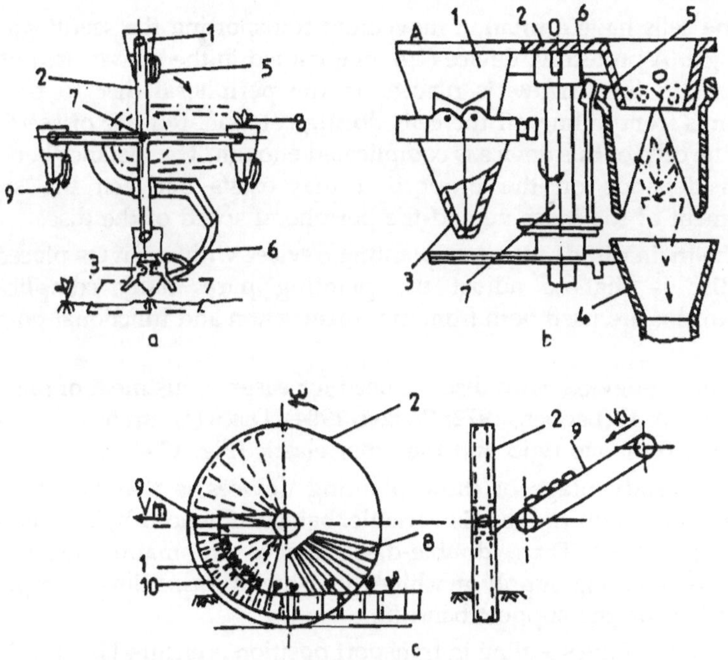

Fig. 19.3 Cup-shaped planting devices fixed on disc.

Inside the cups (1) there are two seamed-cups (3) which release the seedling into the furrow or into the guiding tube (4) rotating round point (0) due to some levers with rolls (5) that come into contact with the cam (6). In this case, the planting device is also activated from the transport wheels or copy wheels which set into motion the horizontal shafts (7) (Fig. 19.3a) or vertical ones (Fig. 19.3b) (Em, 1978; Mikawa, 1983; Scudellaro, 1989).

In the case of cups which have a horizontal rotation, the seedling is transported from the supply point (A) over the guidance tube (4). Vertical transport of the seedling to the furrow opened by the ploughshare is done in a gravitational and pneumatic way using different mechanisms in order to preclude damage to the seedling or changing of its position when entering the soil.

One risk with these devices is damage to the plant-pot during soil intromission or blockage of the tube which leads the seedling from the supply point to the furrow.

To avoid these disadvantages, fixing wings on the vertical disc (2) of the planting device in figure 3.c, whose cups have the shape of cells, is suggested. The seedlings are planted laterally or horizontally by means of a conveyor (9) or manually.

The cells have a rotation movement transporting the seedlings at a 90° angle. A protecting device (10), interrupted in the release area of the seedling in the furrow, is placed on the peripheral disc to prevent seedlings from falling off the cells during rotation. Release of seedlings from the cells of this device is complicated enough. The distance between plants depends on the proportion that exists between the speed movement of the machine and the peripheral speed of the disc.

The main disadvantage of planting devices with receivers placed on the disc is that to adjust the planting pace many complicated mechanisms are used both from the construction and functional point of view.

Planting devices with disc are used for easier adjustment of planting pace (Fig. 19.4) (Lannen, 1972; Pretzer, 1984). Discs (1) can be both elastic (Fig. 19.4.a) or one rigid and the other elastic (Fig. 19.4b, c).

The disadvantage of these planting devices is that the distance between plants on the row is variable, being influenced by incidence of seedling supply. Thus, double-disc devices are mainly used with automatic seedling supply machines in which the seedlings are placed equidistant on the support band (3).

Fixation of the seedling in transport position is ensured by the elastic force of the discs (Fig. 19.4a), the pressing force of some rolls (4) (Fig.

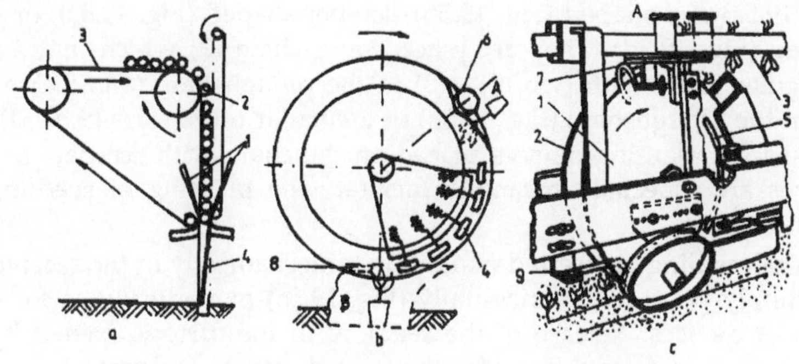

Fig. 19.4 Planting devices with disks

19.4b) and the elastic force of the disc and a cam (5) (Fig. 19.4c). For this, the discs are placed on the inclined shafts at a certain angle so that their peripheral sides come into contact at a certain length.

To insert safely and easily the stem of the seedling between discs in the supply point (A), the discs are moved off by means of a roll (6) (Fig. 19.4b) or a lever (7) (Fig. 19.4c). At the point of releasing the seedlings into the soil the discs are moved off with a roll (8) or a lever (9).

These devices are compact, allowing planting in closely placed rows; they can achieve any distance between plants in the row according to supply frequency and speed movement.

Another group of devices for planting seedlings are those with receivers placed on conveyors with a flexible transport system (Fig. 19.5) (Mugnier, 1986; Tihomirov, 1987). The receivers can be clamp-shaped

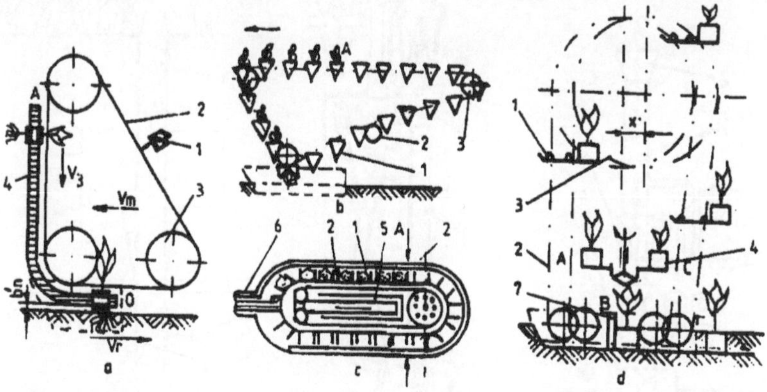

Fig. 19.5 Planting devices with receivers placed on a flexible transport system.

(Fig. 19.5a) cup-shaped (Fig. 19.5b), scraper-shaped (Fig. 19.5c), or shelf-shaped (Fig. 19.5d). They are placed on a chain (2) which may have a triangular route with two rolls (3) in the ploughshare (Fig. 5.a) or one roll in the ploughshare (Fig. 19.5b) or a circular route (Fig. 19.5c, d). The cup-shaped planting devices placed on the chain with scrapers or with shelves are especially meant to facilitate the planting of seedlings in nutritive pots.

The seedling is inserted manually or mechanically in the receivers in the supply point (A), horizontally (Fig. 19.5a) or vertically as for other types of devices. Release of the seedling in the furrow opened by the ploughshare is done differently: the jointed clamp receivers come out of the guiding tube (4) (Fig. 19.5a): the cup-shaped receivers are opened by a cam placed in the ploughshare; in the planting device of scraper-conveyor type there is a pusher (5) with a rectilinear movement which leads each seedling into the drain (6) of the ploughshare and for the shelf-conveyor there is a stopper (7) in the lower part of the route of the ploughshare.

Adjustment of the distance between plants in the row is done by modifying the number of receivers on the conveyor, the speed of the conveyor should be equal to the speed of the machine. The distance between plants in the row can be adjusted to a minimum pace equal to the length of the chain loop.

There are planting devices with gravitational or pneumatic transport of the seedling from the supply point (Fig. 19.6) (Balan, 1998a, b Bumacov and Rosca, 1989; Hung, 1973).

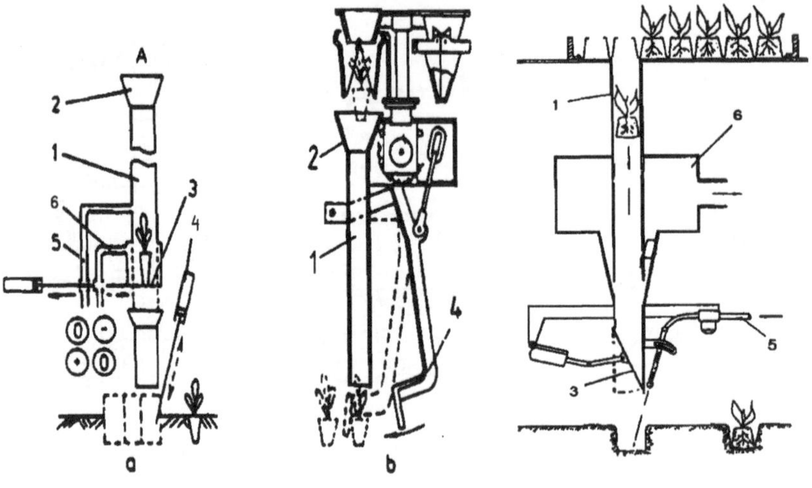

Fig. 19.6 Planting devices with gravitational or pneumatic transport.

The vertical guiding tube (1) has a drain (2) for manual or automatic supply, done directly or by means of a distribution device. These devices are especially meant for seedlings grown in nutritive pots whose weight enables their falling through the tube. A rising air current can be achieved by means of a pneumatic device in order not to block the tube with seedlings. For uniform distribution and to diminish the shock of the plant-pot on impact with the soil, the guiding tube can be closed with a drawer (3), which opens at pre-established intervals (Fig. 19.6a, c). To keep the seedling vertical, even after it reaches the bottom of the furrow, this is supported by an arm (4) which makes a tilting movement.

Planting devices with needles (Fig. 19.7) are exclusively used for seedlings grown in nutritive pots disposed in a certain order in the cells of a container (2) (Fig. 19.7a, b) or on a platform (Branch, 1987; Caponet, 1989; Gaston, 1987; Hakli, 1989).

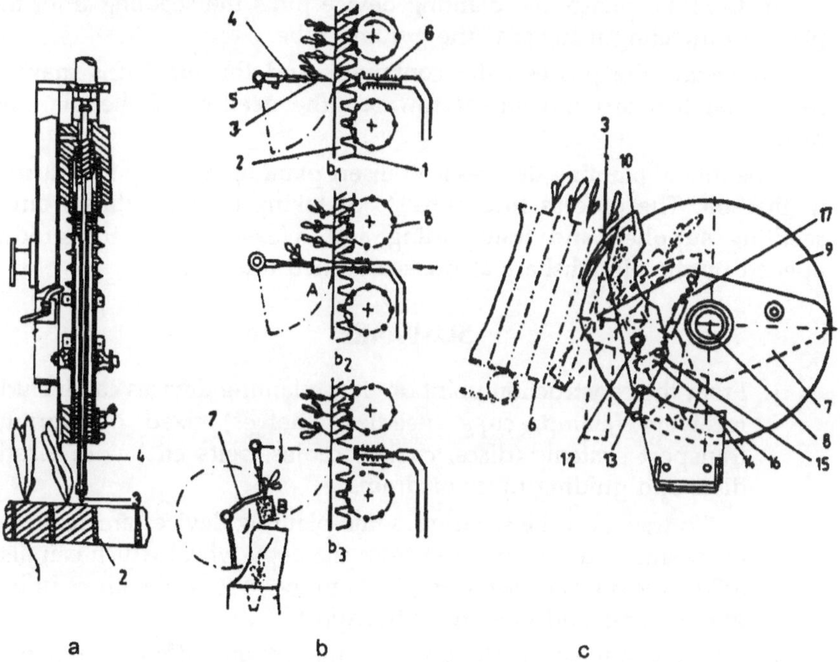

Fig. 19.7 Planting devices with needles.

For the device in Figure 19.7a, the needle (3) can slide inside a bar (4). At first, the needle enters the soil of the plant-pot, fixing the seedling vertically. In the second phase the bar pushes the seedling in the groove opened by the ploughshare, after which there is a withdrawal phase of the needle and then of the bar. The bar and the needle for fixing the seedling make an alternative rectilinear movement activated hydraulically.

The planting device with a needle placed rigidly on the bar (4) in Figure 19.7b makes an oscillation round its axes (5). The seedlings are extracted from the cells of the container (2) by means of a pushing bar (6) which puts the plant-pot in the needle (3). When bar (6) is withdrawn, bar (4) makes an oscillation from point (A) to point (B), from where a revolving pitchfork (7) extracts the pot from the needle (3) directing it vertically in the guiding tube.

The needle (3), fixed on a tilting arm (10) (Fig. 19.7c), makes a complex movement following the profile of the cam (14). The needle, which is in a falling movement, enters the plant-pot, after which it separates the pot from the group, moving backwards.

In the final phase, the planting device pulls the seedling from the platform directing it towards the guiding tube.

To restart the process, the containers and the platforms make a sequential forward movement towards the needles of the planting device.

The role of planting devices is to insert planting material in a furrow in the soil. The process thus consists of taking each seedling from a seedling supplier, and transporting and releasing it in the groove opened by the ploughshare at pre-established intervals.

SUMMARY

1. From the construction point of view, planting devices can include receivers (clamp, cups, needles, shelves) fixed on various transport systems (discs, chains, cables, belts etc.) with elastic discs and guiding tubes or drains.

   To transport the seedling, some planting devices are activated by means of a transmission from the copy wheel which can also adjust the distance between plants in the row, while others utilize gravitational and pneumatic transport.
2. Planting devices with receivers placed on different transport devices carry out uniform distances between plants but this adjustment is limited to the number of receivers and the proportions of the transmission movement. Adjustment of the distances between plants in the row should be done by changing the number of receivers, even if the distances possible are reduced and the adjustment is done only gradually, since to adjust the

proportion of the transmission would negatively influence the value of the work.

3. Planting devices with elastic discs achieve larger possible distances between plants in the row because determined by a regular supply frequency, but the uniformity of distribution depends on uniformity of supply incidence which, in the case of manual supply, is influenced by the experience and degree of exhaustion of the worker; that is why automatic devices are used more often.
4. Planting devices with guiding tubes or drains are exclusively used for seedlings grown in cubes or nutritive plant-pots to level the distances between plants in the row, a process usually done by means of diverse mechanical systems with pre-established action frequency.
5. Planting devices should constantly keep the seedlings in a vertical position, the route of the seedling during the planting process should be shorter and the chances of damaging the seedling or the nutritive plant-pot should be reduced in so far as possible.

References

Balan, O. 1998a. Device for transplanting seedlings. Letters patent no.113922, Romania.

Balan, O. 1998b. Contributions Concerning the working of seedling transplanting machines. PhD. Thesis, The"Gh. Asachi" Technical University, Iasi, Romania.

Branch, L. G. 1987. Method and apparatus for transplanting crops. L.P. no 4644880, USA.

Bumacov, V. and Rosca, V. 1989. Semiautomatic Machine for Seedling Transplanting. L.P. no. 214, Moldova.

Caponet, M. 1989. Machine a planter des vegetaux. L.P. no. 2622763, France.

Ciubarin, M.I. 1972. Rassadoposadocinye masiny. Masinostroenie, Moskva, Russia.

Em, A.D. 1978. Masiny dlia mehanizatii bahcevodsta. Kartofeli i ovosci, nr. 8, Moskva.

Gaston, S. 1987. Nouvelle machine planteuse de planst en mottes. L.P. no. 2299211, Franta.

Hakli, R. 1989. Procede et dispositif pour la plantation de jeunes plants. L.P. no. 2622387, Franta.

Hung, B.K. 1973. Systems engineering at precision automatic transplanting. Trans. ASAE, USA 73: 104.

Lannen, S. 1972. Machine a transplanter les jeunes plants. L.P. no. 2208273, Franta.

Mikawa, I. 1983. Transplanter provided with seedling selector. L.P. no. 58117319, Japan.

Mugnier, F.J. 1986. Element planteur de mottes en culture maraichere. L.P., no. 2569330, Franta.

Pretzer, P.R. 1984. Transplanting apparatus utilising a disc mechanism for transferring plants from a supply tape to the furrow. L.P. no. 4455950, USA.

Scudellaro, L. 1989. Transplanting machine. L.P. no. 4886002, USA.

Tihomirov, A.N. 1987. Opisanie uzobretehnika k avtorskomy sbidetelictou. L.P. no. 1501947, Moskva, Russia.

20

NEW TRANSPLANTING MACHINES FOR SEEDLINGS GROWN IN NUTRITIVE PLANT-POTS

Ovidiu Balan and Vasile Cracium
The "Gh.Asachi" Technical University IASI, Facultatea de Mecanică, Bdul. Mangeron nr.63-65, IASI, Cod. 6600, Romania

INTRODUCTION

Cultivation of crops by transplanting seedlings is especially applicable to vegetables, tobacco, hops, vineyards, fruit trees, some flowers or medicinal herbs, potatoes, bulbs and certain plants for seeds (beet, carrot).

For vegetables and other plants, transplanting seedlings grown in hothouses, plant-pots or greenhouses is an objective necessity due to the following advantages:

- to obtain in constant condition premature plants by harvesting the crop 30-60 days ahead;
- to distribute the production and consumption of fresh vegetables gradually;
- to ensure qualitative sorting of the products, achieving crops with selected and uniform plants;
- to reduce the need for seeds at the surface unit;
- to shorten the period of using the soil, offering the possibility of achieving successive crops and extending the crop area for plants

sensitive to heat and also the possibility of growing such plants in cold regions.

Given these advantages, the world tendency, mainly in well-developed countries with reduced agricultural surface, is to extend this technology to other crops, in particular technical plants.

The seedlings are produced by direct sowing on a nutritive bed, in plant-pots, boxes or cell containers and by sowing or transplanting in plant-pots, boxes or directly in the soil.

To increase plant crops, together with improvement of agricultural machines, it is necessary to use the newest technologies, which suppose cultivation of plants by transplanting seedlings grown in nutritive plant-pots.

The advantages of this technology are: increasing the fixation degree of the seedling (96-98%), continuous growth of the plant, which determines bigger and 10-20 days earlier crops, saving of seedling material, the possibility of complete mechanization of work and increasing the speed of the transplanting machine from 0.25-0.4 to 0.7-0.8 m s^{-1}.

Unlike mechanical planting of simple seedlings for which supply to the transplanting device is done by one hand while taking each seedling from a bunch held in the other, planting of seedlings grown in nutritive plant-pots of different shapes and sizes with supply to the transplanting device done with both hands, taking the seedlings from the place where the plant-pots are placed. In this case the frequency of seedling supply is doubled, reaching 70-90 seedlings min^{-1}.

It is obvious that this increase in work speed and speed of the planting device is not sufficient. That is why new solutions are being sought to increase work speed without a negative influence on the quality of the planting process.

The need for designing new machines is apparent in Romania where the MPR-5 machines are still used for transplanting seedlings in plant-pots.These machines use furrow openers with distant wings and claws which are not very efficient instead of clamps for seedling fixation.

A planting machine with high speed, equipped with group supply of seedlings grown in nutritive plant-pots, has been researched and designed.

MATERIAL AND METHODS

If the supply of planting devices were done in groups of plant-pots the work speed would increase significantly according to the formula:

$$v_m = \frac{n \cdot f \cdot d}{60} [\text{m/s}]$$

where: n is the number of plant pots in the group (4-8);

f the frequency in supplying with groups (30–50 groups per minute); and

d the distance between plants in the row from 0.1 to 1.5 m.

The distance between plants in the row depends on species, environmental conditions and the transplanting machine and cannot be modified by someone's will; this is imposed by agrotechnical conditions and the supply frequency is influenced by subjective factors (worker's experience, fatigue, concentration etc.) and objective factors (constructive parameters of the machine, working conditions). To increase work speed, in this case we can adjust the number of plant-pots in the group (n).

Theoretically, using this method the planting speed can increase at any value; the maximum speed is determined by the value at which the quality of the planting process deteriorates (number of holes increases, regular distribution decreases, deterioration degree increases etc.)

The first condition to make these machines to work is that the seedling supply should be distributed on a platform, from which the worker can easily take a group of plant-pots and place them in the distribution device of the machine.

For this purpose a container (8) was designed in which conical cells are placed in a matrix (line and columns) (Fig. 20.1) (Balan, 1998).

To transfer the group of seedlings from the container to the transplanting device, rigid plastic, wood or metal bars (26) are juxtaposed on the container.

The bars are cell-shaped on the edge and form extensions. The bars have a handle (27) at one end and a catching hook (b) and hanger (a) at the other.

By joining the two bars, by means of hooks, a number of plant-pots are fixed between bars which are taken from the cells from the upper side and transferred to the planting device.

The planting device is a modular type on a frame (1) with one or more working sections (A), a water tank (2), a triangular attachment (3) to the suspension system of the tractors and a bower (4).

A working section (A) (Fig. 20.2) consists of an upper table (5), fixed frame (1) and lower frame (6), bent on the front side of the table (5) by means of a support.

A distribution device (7), two containers (8) with seedlings, a support (9) for two other containers with seedling supply and a tube (10)

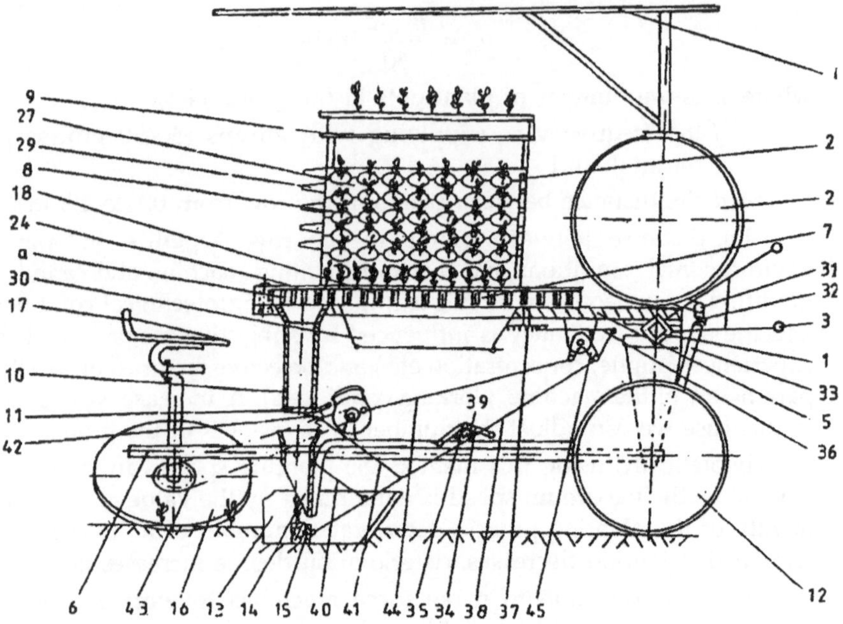

Fig. 20.1 Side view of the transplanting machine.

for guiding the seedling which has a clamp (11) inside are placed on the upper side. On the lower side (7) there are a coping wheel (12), a furrow opener (13), a divided tube (14) to sustain and protect the stem, a support plate (15) for the plant-pot, two pressing wheels (16) and a chair (17) for the worker.

The distribution device (7) has the role of transporting the group of seedlings and distributing each in the guiding tube (10).

Device (7), detailed in Figure 20.2, consists of a chain conveyor (18), table with nutritive plant-pots (5) and a closed circuit.

L-shaped elements (19) fixed at regular distances on the outside of the chain–conveyor form a succession of U-shaped compartments (20), where the nutritive plant-pots are placed. The compartments have a margin (21) on the outside in the transport area. The conveyor (18) is activated by a pressing and stimulating oscillating lever, having irregular movements corresponding to the space covered equal to the pace between two compartments.

The distance between the components is equal to the distance between the cells on the bar length and the number of cells between two bars (a row on the container) is a submultiple of the number of compartments on the conveyor in the proportion of 1:3 or 1:4.

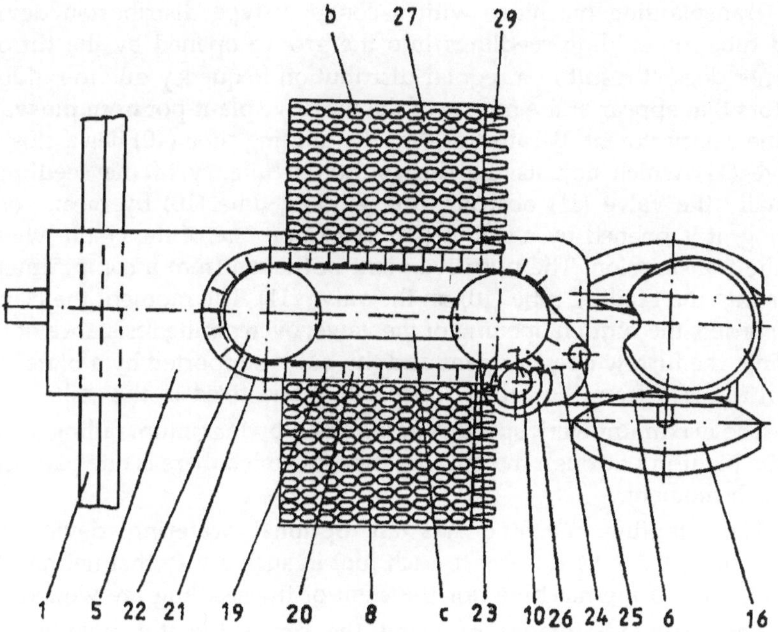

Fig. 20.2 Top view of the transplanting machine (without bower or tank).

Each group of compartments has the extremities marked (for example, in green and red) in such a way that between two successive supplies on each side of the conveyor there should be between the two ends a number of compartments equal to the number of seedlings on a row.

For automatic extraction of the plant-pots from the transport compartments, the outside of the conveyor is wrapped on three sides with a continuous band (22), which has some equidistant windows through which the separating walls of the compartments extend. Thus, the conveyor also activates the band (22) which over the guiding tube (10) extends farther from the conveyor because of a stretching roll (23), passing by a scraper (24) which makes the plant-pots fall into the guiding tube.

On each side of the distribution device there is a container fixed in a sloping position to the table (5) where the nutritive plant-pots are placed. The worker takes two bars, joints them, lifts the plant-pots placed between bars from the cells and moves them to one side of the conveyor with empty compartments where, by moving the bars away, the group of plant-pots are released one by one in each compartment.

Transplanting machines with a conveyor-type distribution device and tube for guiding seedlings into the groove opened by the furrow opener doesn't result in a regular distribution frequency due to random factors that appear at the release of the nutritive plant-pot from the walls of the compartment. Because of this, the guiding tube (10) has a closing valve (11) which adjusts the distribution frequency of the seedlings. Usually the valve (11) obstructs the guiding tube (10) by means of a spring; it is opened by a clamp (25) placed on one of the chain wheels of the transmission. The nutritive plant pot taken from a compartment, falls into the guiding tube (10) on the valve (11). The moment the clamp (25) orders the sudden opening of the valve, overcoming resistance of the spring, the furrow opener opens and the pot is supported by a plate (15) until released from the furrow opener (13) and fixed in the soil.

The maximum work speed is determined by maximum falling speed of the plant-pot through the guiding tube, in which there is one plant-pot at each moment.

The planting machine has an optional watering device for distributing 0.2 – 0.4 L water to each root in such a way that neither the metal parts of the machine nor the stem of the seedling are wetted.

The main advantages of using the semiautomatic machine for transplanting seedlings by supplying the distribution device with groups of seedlings are:

- introduction of the most modern technologies for growing crops by planting seedlings grown in nutritive plant-pots;
- increase in work speed by supplying the distribution device with groups of nutritive plant-pots.

The main disadvantages of the semiautomatic machines for transplanting seedlings are the low work speed and the need for manpower: a tractor driver, a worker for each section and another two or three auxiliary workers to transport and supply the planting machines with boxes of seedling bunches.

To eliminate these disadvantages, a semiautomatic machine for seedling tramsplantation was designed, that excluded worker-operators for sections, augmented work speed, reduced machine weight and consumed less fuel due to increase in working speed and decrease in working machine resistance (Balan, 1999).

This redesigned transplanting machine (Figs. 20.3 and 20.4) is mounted on the rear of a tractor. The frame (1) can carry more working sections, corresponding to the number of rows to be planted at one pass, one or two water tanks (2) and two wheels (3) which copy the surface of the soil. These wheels (3) drive the shaft (4) mounted on the frame (1).

The machine consists of several modules. Each section consists of an upper table (5) with a frame (6) beneath it, mounted on frame (1). A seedling-input device (7), disc (8) for catching emptied seedling cells, and guiding tube (9) for seedling with valve (10) for adjusting planting interval are mounted on table (5). The furrow opener (11), two seedling wheels (12), cut tube (13) for support and protection of seedling, and pot-supporting plate (14) are mounted on the lower frame. Unevenness of the ground is copied a stick (15) with spring (16) between the lower frame and the upper table. The seedlings are located in alveoli (cells) between two elastic bands (17) with shoulders (18) fronting each other. The two bands are tied in a spiral around a disc (19).

Seedlings are distributed by driving the bands over the tube (9) where the pot cells open. The nutritive pots fall from the open cells onto the valve (10); the pots fall at regular intervals into the ditch made by the furrow opener (each pot is supported until its emergence from the furrow opener) and are fixed in the ground by the plate (14) and tube (13). Seedlings are watered from tank (2) by tube (20) and the 3-armed tap (21), driven by a lever system (oscillating lever (22)), crank and connecting rod assembly (23), from the driving mechanism.

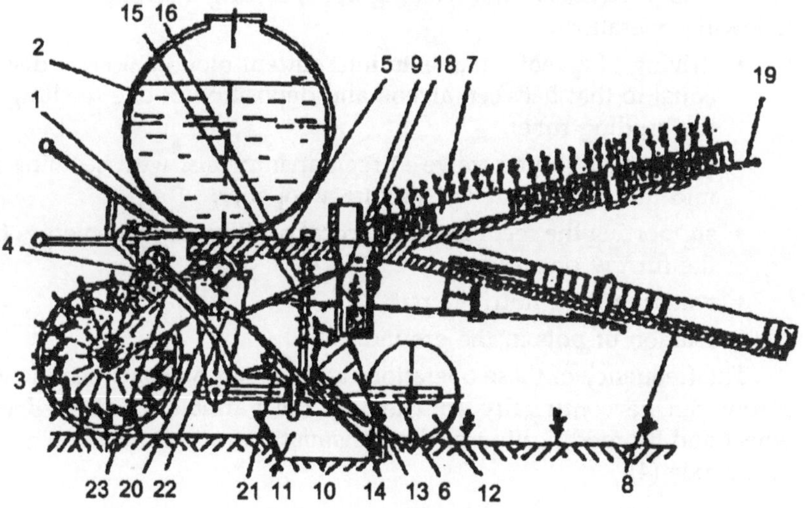

Fig. 20.3 Automatic machine for transplanting seedlings.

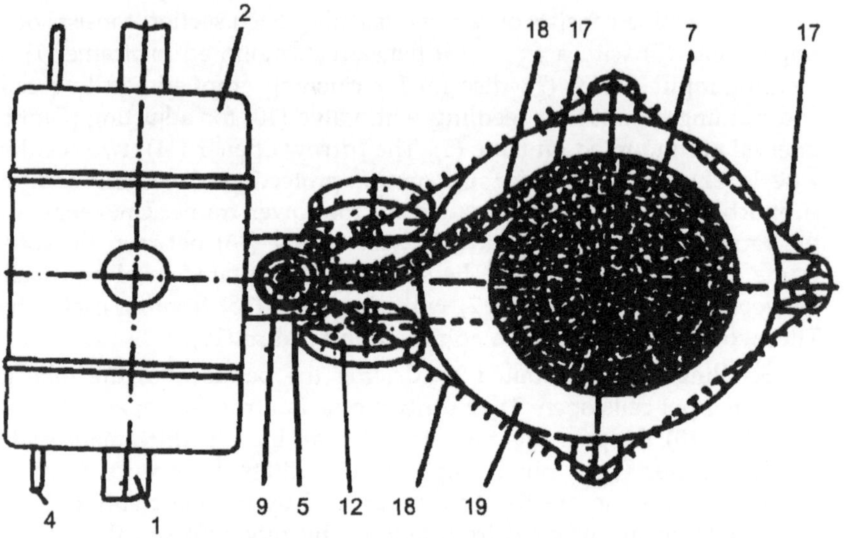

Fig. 20.4 Working machine module (top view).

The working process of the seedling transplanting machine involves the following operations:

- driving of alveolar bands in intermittent motion along a distance equal to that between alveoli and deposition of one seedling into the guiding tube;
- opening of shutter valve at regular intervals, with seedling sent into the ditch made by the furrow opener;
- supporting the seedling in a vertical position until it emerges from the furrow opener;
- watering each nutritive pot;
- fixation of pots in the ground.

The frequency of these operations and the interval between plants in a row can be continually adjusted by the variable-diameter driving wheel and by modification of the transmission ratio from the wheel (3) to the axle (4).

CONCLUSION

The constructive and functional improvement of the seedling transplanting machines presented above, assure improvement of plant quality, increase in work speed and work productivity, decrease in labour force, increase in production per hectare and decrease in fuel consumption.

References

Balan, O. 1998. Contributions, concerning the working of seedling transplanting machines. Ph. D. Thesis, The 'Gh. Asachi' Technical University, Iasi, Romania.

Balan, O. 1999. Automatic machine for transplanting seedlings grown in nutritive pots. Patent No. 113923, Romania.

21

THEORETICAL CONSIDERATIONS AND SOME EXPERIMENTAL RESULTS CONCERNING IMPROVEMENTS IN SEEDLING TRANSPLANTING MACHINES

Vasile Cracium[1], Vasile Bumacov[2], Ovidiu Balan[1] and Ilie Filipescu[3]
[1]The 'Gh. Asachi' Technical University IASI, ROMANIA
[2]The Agrarian State University CHISINAU, Republic of Moldova
[3]The 'Agromec' S.A. Movileni, Iasi county, Romania

INTRODUCTION

Manual transplanting of seedlings require considerable time and labour expenditure and is extremely tiring. Today, in Romania, most vegetable cultivation is done in small plots and seedlings transplanted manually (about 6-8 pieces per minute).

With mechanized transplantation, about 43-48 seedlings can be planted per minute in one row, i.e. 5-6 times more than by manual transplanting (Bernacki, 1972). In Romania and Moldova for mechanized transplantation Romanian-made machines are mostly used, from the series MPR-5(6, 8) (1987) and only a few imported machines (John Deere, Super Prefer, Piccard etc.). The transplanting machines are, for the most part, semiautomatic. Undoubtedly, mechanized transplanting of seedlings has many advantages over manual transplanting but there are some disadvantages too, namely:

- many seedlings are fixed in the soil with a deviation to the vertical;
- seedling roots are naked, without soil, and hence the seedlings suffer damage when fixed in machine grabs;
- seedlings picked a hothouse tend to wither until transplanted into the humid soil mass and suffer a few days retardation (Hung, 1973), until their roots begin to grow (i.e., vegetation starts up again).

To avoid the aforesaid disadvantages, one solution is *use of seedlings grown in nutritive pots and transplanted in the field, using specially adapted semiautomatic machines*. The technologies of growing seedlings in nutritive pots and new semiautomatic transplanting machines give far better qualitative indices than the old technologies and machines.

SOME THEORETICAL CONSIDERATIONS

Due to the fact that the *transplanting devices of the machine are manually fed with seedlings by a human operator, the transplanting machines are designated semiautomatic* (Fig. 21.1).

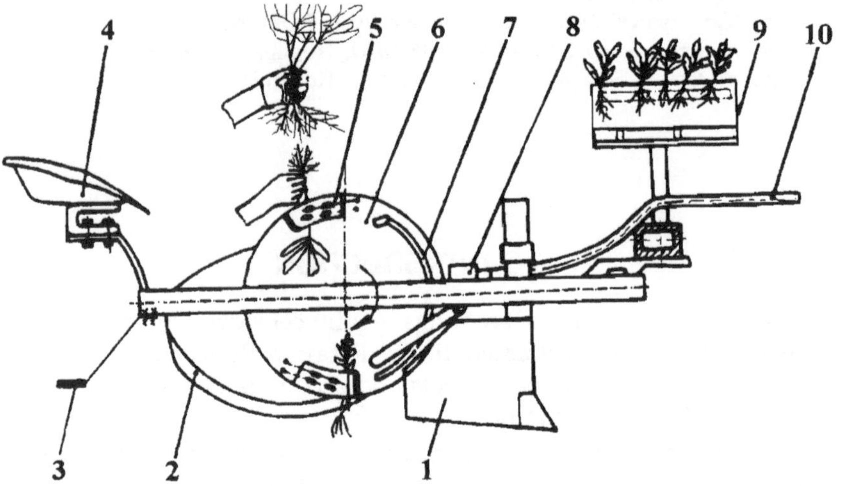

Fig. 21.1 Technological scheme for transplanting seedlings with the machine MPR-5(6) (Romanian make). 1-Furrow opener; 2-press wheel; 3-levelling bar; 4-worker's seat; 5-grabs; 6-disc; 7-cam rod; 8-cock; 9-seedling box; 10-water pipe (from 'Transplanting Machine' MPR-5...).

A common transplanting machine consists of individual sections fitted on a frame. Each section represents a single-row complete machine. The seedlings are located in containers (9) on the platforms fitted on the machine frame. Seated on the saddle the operators (4) take a seedling from the container (9) and place it roots downward on the open surface of the revolving grab (5). The seedlings are conducted to the soil and upon release from the holder immediately enclosed with soil by two press wheels (2) obliquely positioned to each other. The revolving grabs (generally, the transplanting device) are driven by one of the press wheels (2), through a gear transmission at the usual ratio of i = 1 – 1.5. It follows from the foregoing that the speed and productivity of the transplanting machine depend on the ability of the operators who feed the grabs on the transplanting disc. Experimental researchers (Bernacki, 1972; Bumacov, 1997; Balan, 1998) have shown that in the field, the frequency of feeding (depending on operational ability) is 43-48 seedlings per minute, which means an average time of about 1.30 seconds for each plant.

Taking into consideration a transmission gear ratio, i = 1, to ensure the vertical position of seedlings, the speed of the machine v_m is:

$$v_m = \frac{R_{grab} \cdot \varphi}{t} \cdot (1 - \delta) \qquad (1)$$

where R_{grab} is the radius of revolution of the grab centre (250-270) mm;

$\varphi = \frac{2\pi}{z}$, the angle between two consecutive grabs on the disc;

z the number of grabs on the disc;

t the average feeding time (about 40 seedlings min^{-1}, $t \cong 1.5$ s.);

δ the tractor wheel slippage (0.1–0.2).

Assuming $R = 300$ mm, $\varphi = \pi/2$ (equivalent for 4 grabs on the disc), $\delta = 0.11$ and $t = 1.3$ s, yields $v_m = 0.3226$ ms^{-1} = 1.16 km h^{-1}.

It is clear from the above that the operational speed of the transplanting machine depends on the reflexes (ability) of the operators who insert seedlings into the grabs (or between the discs) and for this reason the speed is relatively small (0.8–1.4) km h^{-1}.

Laboratory and field tests undertaken to increase labour productivity and machine speed as well as improve working conditions for the operators, indicated two major directions for achieving these goals, namely:

a. Improve technological schemes of the semiautomatic transplanting machines for seedlings grown in nutritive pots

(Bumacov and Rosca 1995; Bumacov, 1997; Hung, 1973; Moden et al., 1977; Balan, 1998);

b. Automate feeding of the transplanting device (Balan, 1998).

The authors consider that method *a* meets the conditions encountered by farmers due to its advantages. Method *b* gives good results but is not succesful for the situation common in Romania and in the majority of the countries with a similar economic profile.

Figure 21.2 shows a general view of a semiautomatic transplanting machine for seedlings grown in the nutritive pots and Figure 21.3 presents the kinematics scheme for the machine depicted in Fig. 21.2 (Bumacov and Rosca, 1995). The machine presented in Figure 21.2 consists of individual sections connected by a common frame, mounted on a tractor. Each section represents a single-row complete machine.

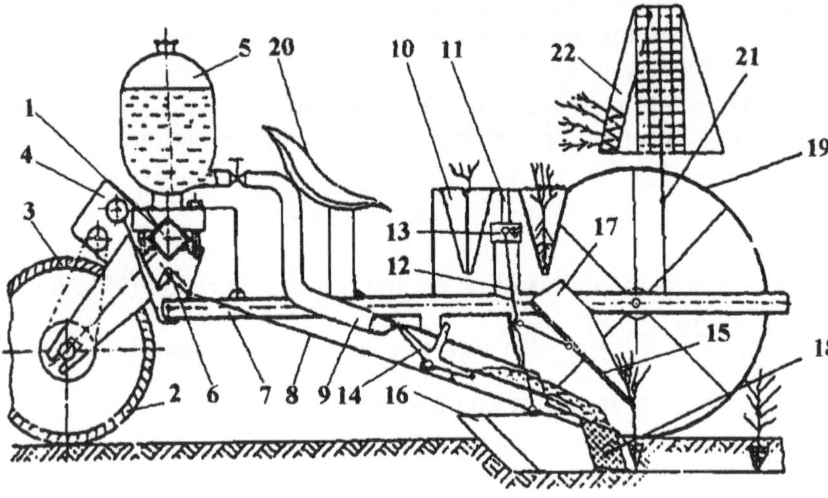

Fig. 21.2 Semiautomatic transplanting machine for seedlings grown in nutritive pots.

The concept is that the seedling with its nutritive pot must drop into the open furrow in a vertical position, after which it is ejected from the planting shoe at the same velocity as the forward travel of the machine, but in the opposite direction.

For synchronization of working operations and maintenance of constancy of spacing of plants in a row independent of machine speed, all the mechanisms on the machine are driven by the driving wheel (2). The only mission for the operator, seated on the saddle (20) is to remove the seedlings from the germination flat (cartridge) (22) and place them in the carousel

(11) (which rotates in the horizontal plane), which is fitted with magazine retainers (10). When the magazine retainer 10 reaches the point over the drop tube (17), the retainer releases the seedling which drops into the open furrow, in front of the ejector (18), which in turn pushes the seedling and sustains it in a near-vertical position. Packing wheels (19) cover the seedling roots with soil at the desired compression. On the platform is located a water container (5) for automatic watering of the seedlings; water flows out through a special distributor (9) driven by the same mechanism which commands the ejector (18). Obviously, as stated above, it is essential to choose the proper driving mechanism for the transplanting machine and to design the simultaneous phases for the transplanting process.

According to the technological scheme of the transplanting machine shown in Fig. 21.2, the ejector CD pushes the pot from F to H, with a stroke "s". From Figure 21.2, the value of stroke "s" is obtained as follows:

$$s = 2.l.\cos\alpha_{max} = 2.l.\cos\frac{\sqrt{e^2 - r^2}}{e} \qquad (2)$$

where l is the length of the AC bar (m); α_{max} the maximum value of angle α;

- α the instant value of angle of the bar AC measured to the vertical position AO;
- r the radius of the handle OB (m); and
- e the constant distance between centres of rotation O and A (m).

The value for α_{max} is given by the formula:

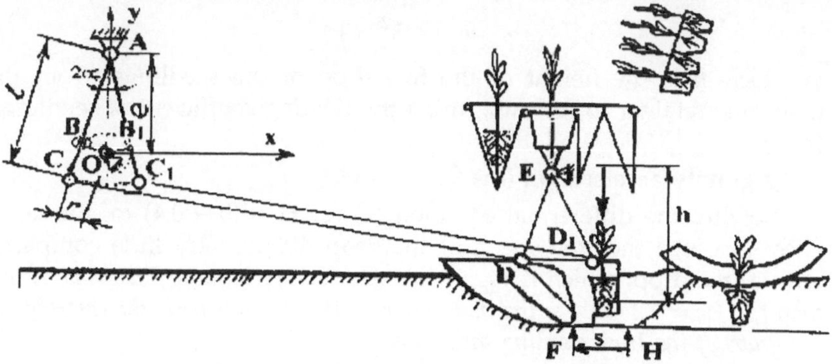

Fig. 21.3 Kinematics scheme of the machine shown in Figure 21.2.

$$\alpha_{max} = \text{arc sin } \alpha_{max} = \text{arc sin}\frac{r}{e} = \text{arc cos } \alpha_{max} = \text{arc cos}\frac{\sqrt{e^2 - r^2}}{e} \quad (3)$$

To ensure a quite rectilinear movement of point F of ejector CDF, we can consider the mechanism ACDE as a parallelogram, which is driven by the handle with arm OB.

When the ejector moves from F to H, the average speed is:

$$v_{push} = \frac{FH}{t_{push}} = \frac{2l.\cos\alpha_{max}}{\frac{\pi + 2\alpha}{\omega}} = 2l\omega.\frac{1}{\pi + 2\alpha}.\cos\frac{\sqrt{e^2 - r^2}}{e} \text{ (ms)} \quad (4)$$

$$v_{pull} = \frac{FH}{t_{push}} = \frac{2l.\cos\alpha_{max}}{\frac{\pi - 2\alpha}{\omega}} = 2l\omega.\frac{1}{\pi - 2\alpha}.\cos\frac{\sqrt{e^2 - r^2}}{e} \quad (5)$$

where, ω (s^{-1}) is the angular speed of the handle OB.

From the above it is evident that $t_{push} > t_{pull}$. For a correct function, t_{drop} of the seedlings through the tube (17) (Fig. 21.2) must be less than t_{push} and concomitantly less than the time between the consecutive feeding of the cartridge by the operator, time which is marked as $t_{min.min}$. This time is a *physiological time* and is given by the operator's ability. Experimental researches have indicated that $t_{min.min}$ is:

$$t_{min.min} = \frac{60\,(seconds\,/\,minute)}{43....50\,(seedlings\,/\,minute)} = (1.4 - 1.6) \text{ sec/seedling} \quad (6)$$

The drop time of seedlings through the tube (17), t_{drop}, can be calculated with the differential equation (7) (Bumacov, 1997):

$$h + k.h^2 = g \quad (7)$$

in which h is the height of the free drop of the seedlings (from the magazine retainer to furrow), (m); k the floating coefficient of seedlings, (m^{-1}); and

g gravity acceleration (ms^{-2}).

Solving the differential equation (7) for H = (0 – 0.4) m, Bumacov (1997) showed that the speed of the drop differs, very little compared with a free drop, which is t_{drop} = (0.285 – 0.35) seconds. *This time is less than $t_{min.min}$ = 1.2 – 1.4 s, the time in which the operator feeds the cartridge or the grabs of the transplanting disc.*

The machine for transplanting seedlings grown in nutritive pots exhibits some important peculiarities. The operator makes only two simple operations, namely:

a. *takes the seedling from the container (1, Fig. 21.2)* and

b. *puts the seedling in a normal position into the retainer (2), of the cartridge (3), which stops when it is in position over the tube (17).*

Because of these facts, the conditions for feeding the cartridge by the operator are easier compared to those for setting seedlings into the grabs of the transplanting disc. That is why the frequency of feeding increases to over 80 seedlings per minute. In these conditions, the feeding time for a seedling is:

$$T = 60 \text{ (seconds/min)}/80 \text{ (seedlings/min)} = 0.75 \text{ s/seedling} \quad (8)$$

Consequently, for a = 20 cm row spacing, the speed of the transplanting machine, v_{20} will be:

$$v_{20} = a/T = 0.2\text{m}/0.75 \text{ s} = 0.266 \text{ m/s} = 0.96 \text{ km h}^{-1}, \quad (9)$$

and for a = 50 cm row spacing, the machine speed is:

$$v_{50} = a/T = 0.5\text{m}/0.75 \text{ s} = 0.666 \text{ m/s} = 2.4 \text{ km h}^{-1}. \quad (10)$$

These speeds are two times faster than the speed of transplanting machines with grabs on discs (or with elastic discs).

EXPERIMENTAL RESULTS

Experimental tests took place in a laboratory stand in which a machine with 2 sections was tested, and in the field with machines with 4 and 6 sections (Bumacov, 1997). Cabbage seedlings were transplanted in the field and the data obtained are presented in Table 21.1.

Table 21.1 Performance of seedlings (grown in nutritive pots) transplanter

Transplanting machine requirements	*Values of qualitative indices*	
	Demanded	*Obtained*
Operational speed (m/s)	0.5	0.86
Interrow spacing	90	90
Row spacing (pieces/m)	0.83...6	2.5
Required depth (cm)	6-12	8.2
Seedlings at required depth (%)		97.5
Faulty planting (%)		3.8
Position to the vertical (%); to 30%	Min. 80	80.7
— from 30% to 45%	Max. 15	13.3
— over 45%	Max.. 5	6
Row spacing (cm)	40	41.2
Damaged plants (%)	2	0.7
Automatic watering (L/seedling)	0.2...0.35	0.36
Drawing out force (N),	2	6
Specific draft force (kN/m)		1.6

It can be seen from the Table 21.1 results that the qualitative indices obtained with this machine using seedlings grown in nutritive pots meet the requirements imposed by the farmers. Besides that, better working conditions for the operator are assured.

CONCLUSION

Seedlings grown in nutritive pots proved to be a workable system for vegetable cultivation. This system of growing seedlings has some major advantages over the old system, namely:

1. Seedlings grown in nutritive pots have a much greater mechanical and biological resistance; they do not suffer the 10-day interval of removal from the hothouse to transplantation in the field. Once introduced into the soil they grow roots fairly soon, growing is continuous and the yield appears two weeks earlier than in the old system of transplanting. This fact brings important economic advantages.
2. Seedlings grown in nutritive pots are more suitable for mechanized transplanting, using machines such as the one shown in Figure 21.2. The single mission of the operator is to fill the carousel with seedlings taken from the cartridge (60-80 seedlings/ min^{-1}).
3. The ejector incorporated in the transplanting mechanisms has major importance in placing the seedlings in a vertical position in the furrow and in obtaining very good qualitative indices for the transplanting process.

References

Balan, O. 1998. Contributions concerning the working of seedling transplanting Machines. Ph.D. thesis, The 'Gheorghe Asachi' Technical University, IASI, Romania.

Bernacki, H. et al., 1972. Agricultural Machines, Theory and Construction Vol. I. NTIS, U.S. Dept. of Commerce, Springfield, V USA.

Bumacov, V. and Rosca, V. 1995. Semiautomatic Machine for seedling transplanting. Letters Patent, Republic of Moldova.

Bumacov, V. 1997. Theoretic basis for manufacturing and improvements of technologies for sowing and transplanting agricultural crops. Ph. D. thesis, Chisinau, Republic of Moldova.

Bumacov, V. and Rosca, V. 1995. Modern Technologies for Growing Seedlings. UASM, Chisinau, vol. 3, pp. 105-108.

Hung, B.K. 1973. Systems Engineering at Precision Automatic Transplanting. Trans. ASAE, paper nr. 73.

Moden, L. Walter et al., 1977. A mechanically containerized seedling transplanter. Trans. ASAE 20 (1): 38-40.

Transplanting Machine MPR-5. Ministry for Agriculture and Forestry; Technical Information, 1987. Bucuresti, Romania.

22

FRUIT PRODUCTION QUALITY IN LATVIA

Maija Eihe
Latvian State Centre of Plant Protection, Latvia
e-mail: iscpp@pąrks.lv

Fruit growing in Latvia is recognized as a prospective branch of agriculture. The total orchard/fruit area used to be about 12,000 ha, but this value has changed recently because old fruit trees and berry bushes are removed and new, more intensive ones planted. It is calculated that fruit-growing areas in Latvia could be 50,000 ha. Latvian fruits are distinguished by a very good bouquet and aroma compared with those imported from more southern regions. The moderate climatic conditions make qualitative fruit production possible with comparably low use of chemical plant-protection products. Minimal use of chemicals has been the main task of plant-protection investigations in orchards in many years. Use of chemicals on flowering plants was forbidden. Our institution, presently LSCPP, elaborated the basic plant-protection schedule for apple orchards with three fungicidal applications per season with a footnote that additional treatments would be necessary if apple scab-promoting conditions set in. The basic index for the first fungicidal application was the beginning of apple scab ascospore discharge. This commonly occurs in late April or early May. Such a schedule sufficed for old, scab-resistant apple varieties widespread in Latvia at this time. Moreover, investigations were carried out in the comparably dry southern region of Latvia. Recently the assortment of varieties has enlarged, including more qualitative but simultaneously more scab-susceptible ones. The previously elaborated plant-protection schedule was insufficient. Taking into consideration our knowledge of the critical primary scab infection periods, we considered measures for a scab warning system and followup of ascospore discharge throughout the entire primary scab infection period. Subsequently, having introduced the apple scab warning system in Latvia, the aim was not to decrease use of plant-protection products, as in the majority of apple-growing countries, but to determine corresponding application terms. At present we have no scab-warning equipment and determine the critical scab infection periods by the method of vaseline-smeared slides, counting released ascospores after every rain. The most dangerous scab infection periods are those that set in during the apple tree flowering period: unprotected flowers get infected and all the harvest can be lost. In this instance fungicidal use on apple trees during flowering as permissible. Use of fungicides is also allowed on flowering strawberries, mostly with Euparen (tolyfluanid), but is not prescribed for other fruit and berry crops; use of most plant-protection products is limited during the period from flowering until harvesting of berries and cherries.

The main task of LSCPP research work is currently efficacy evaluation of plant-protection products for registration in the country. Horticulture trials of fungicides and insecticides are mostly undertaken in apple orchards and to a lesser extent in strawberry, black currant and cherry plantings. A total of 13 fungicides and 11 insecticides were registered for fruit crops in Latvia in 2001. From 1997 trials with apple trees focused on the highly scab-susceptible variety Lobo under fungal disease promoting conditions—near the Baltic Sea—for more precise evaluation of fungicidal efficacy. Under such conditions four to five applications of fungicides were necessary for satisfactory control of apple scab and other diseases. Protective fungicides such as Champion (Cu hydroxide) 50 WP, Effector (dithianon) WG, Euparen-M (tolyfluanid) WG and curative product Chorus (cyprodinil) 75 WG were tested and compared with the standard Dithane M-45 (mancozeb) WP and untreated control. The fungicides were tested in several doses but in all cases proved more effective at higher doses. The aforesaid products were applied mainly in the first protective application and later in the vegetation season, during secondary scab infection. During the primary scab infection period the curative fungicide Score (difenoconazole) 250 EC 200 ml ha^{-1} was commonly used. Chorus at 300 g ha^{-1} proved highly effective in the first application but it is necessary to take into account its comparably shorter length of post-application effect. If the next application was delayed, with Chorus treatment scab on the apple leaves spread more intensively than in other treatments. In summer treatments for scab control on the fruits, Chorus was moderately effective but considerably decreased apple bitter rot extent in storage. Champion at 4.5 kg ha^{-1} decreased the infected surface of leaves twice better compared to Dithane at 2.4 kg ha^{-1} given the high scab infection pressure during the first protective application. Such decreased scab infection continued not only on leaves throughout the vegetation season, but also on fruits until harvest subsequent to uniform applications with Score. By moderate scab infection pressure at the time of application Champion and Dithane were similarly effective. Euparen-M at 3.0 kg ha^{-1} was significantly more effective compared with Dithane in summer applications (the last time 03.08.) in controlling scab and brown rot of fruits. Effector at 0.6 kg ha^{-1} was similar to Dithane in control of apple scab but more effective in decreasing spread of fruit brown rot.

All the fungicidal applications significantly ensured standard quality fruit production: 75–95% marketable quality fruits from the total harvest versus 7% from untreated areas or no standard fruits or no harvest as such. Effector was tested for efficacy in control of cherry leaf drop (*Blumeriella jaapii*) and also black and red currant leaf spot (*Pseudopeziza ribis*). It proved effective if applied before flowering or at the first symptoms of disease post-harvest. The efficacy of Ronilan (vinclozolin) WG (1996, 1998) and Euparen-M (1998, 2000) was tested in strawberry plantings for control of grey mould (*Botrytis cinerea*) of berries. In all cases the amount of rotten fruits decreased 1.3 to 3 times while yield increased 30 to 47%. Results of the trials proved how very important it is to consider weather conditions vis-a-vis fungicidal applications. There was no difference in results between one and two applications with Ronilan if between both treatments dry weather conditions prevailed. In 2000 sufficient effectiveness of Euparen was obtained with one treatment at the beginning of flowering. In former trials in Latvia with Euparen a 157% higher yield or no effect was obtained dependening on weather conditions during applications. For control of *Verticillium* wilt and other root and crown diseases only use of biological plant-protection products is recommended; soil disinfection is forbidden. The Latvian company 'Bioefekts' produces Trichodermin B-J, D and SC, Biomix B-J, D and Bio B-J 1–5, containing *Trichoderma* spp. and other microorganisms: *Streptomyces, Pseudomonas, Azotobacter, Rhizobium, Polyangium* spp. etc., which are recommended as antagonists against grey mould, root rots, mildew and downy mildew of strawberries, vegetables and ornamentals.

23

REVIEW AND PROSPECTS OF VEGETABLE PRODUCTION IN SHANGHAI

Weimin Zhu and Meilian Yu
Shanghai Academy of Agricultural Sciences, Horticulture Research Institute, Shanghai key Laboratory of Protected Horticulture Technology,
2901 Beidi Road, Shanghai 201106, China
e-mail: saasdic@online.sh.cn

INTRODUCTION

Statistics at the end of October showed that 2000 the area under vegetables was more than 14 million hectares throughout the country, with an output value of 300 billion Chinese yuan, just second to the output value of grain crops. The planting area of vegetables is only 10% of the total planting area in China but creates 20-25% of the total output value of the agricultural sector.

Shanghai is situated on the tip of the Yangtse River Delta and the midpoint of China's coastal line. With a pleasant northern subtropical monsoon climate, Shanghai has a reasonably good environment for agricultural production. At present, the area under vegetables in Shanghai is about 40,000 ha, of which about 4000 ha are under protected culture. As the largest city in China, with a population of 14 million, Shanghai has become one of the largest markets for vegetable consumption in the country. Annual sales of vegetables is 2.4 million tons which is 4500 tons daily on average, with more than 50 species available year around. Any change in the vegetable consumption market in Shanghai would have a direct impact on vegetable production in neighbouring provinces and others which use Shanghai as the main market for their produce.

In the past, vegetables were short in supply in Shanghai and many other areas in China, especially during offseasons. But now, thanks to financial and technical input, especially after three rounds of implementation of the 'Vegetable Basket Project' in Shanghai, more vegetables of good quality are not only supplied to the local market, but also to other places in China; some vegetables have even entered the international market. Vegetable production in Shanghai has thus become an important part of vegetable production in China, especially in the area of the middle and lower reaches of the Yangtse River.

PAST AND CURRENT SITUATION OF VEGETABLE PRODUCTION IN SHANGHAI

Diversified Vegetable Species in Shanghai

Development of vegetable species in Shanghai can be divided into five periods starting from the sixteenth century.

First period: 16th century to early 19th

Most cultivated vegetables were local traditional species, such as Chinese cabbage, mustard, spinach and other leafy vegetables. Root vegetables included radish, carrot and others; legumes green soybean, broad-bean, pea; gourds, mainly winter gourd and cucumber; water shield, *Gordon euryale,* common cat's-tail and other hydrophilous vegetables. Wild vegetables comprised about 50 kinds, such as brake and wormwood.

Second period: 19th century to early 20th

There were many more leafy vegetables and fewer fruit vegetables. Root vegetables and hydrophilous vegetables decreased. Tomato, cauliflower and other vegetables from foreign countries were first introduced in China. Vegetable species cultivated in Shanghai rose to 55.

Third period: 1920s to 1940s

This period can be termed the 'galaxy of Chinese and foreign varieties, and gathering of south and north varieties'. Vegetable species increased remarkably. Local traditional vegetable varieties could be cultivated year round. Tomato and cauliflower were widely cultivated in China. Artichoke, broccoli, dill and other vegetables were introduced in China during this period. Vegetable species increased to about 70 kinds.

Fourth period: 1950s to 1980s

Vegetable variety introduction, selection and breeding were systematically carried out in these years. A number of vegetable varieties were bred in the 1970s, especially hybrids of tomato and cabbage. A total of 88 species of vegetables and over 320 varieties were cultivated in Shanghai, according to the 'Shanghai Annals of Vegetable Variety' in 1959. During this period, Shanghai attained first place in China in carrying out systematic genetic improvement of vegetables and gained a reputation in vegetable production for 'rich species, pioneering research and good variety'. The varieties of Pak-choi, cabbage and tomato from Shanghai were well known throughout China. Cauliflower, cucumber, broccoli, sweet pepper, egg-plant and other vegetable varieties were also widely welcomed throughout China. Lettuce, sweet corn and asparagus were introduced from foreign countries. In the Shanghai area, every district and county had a vegetable production area with distinctive features.

Fifth Period: 1980s to the present

New vegetable species and varieties have been introduced and more and more vegetable varieties are bred and used in production. Traditional vegetable varieties are developed to meet the demand. There are around 120 vegetable species available on the market. The quality of vegetable varieties has improved and species are diversified.

Protected Culture of Vegetables Developed and Yield and Economic Efficiency Improved

Shanghai was the first place in China to practice vegetable protected culture. It plays a leading role in the country in terms of level of protected culture, management experience and technology. A study on integrated vegetable cultivation techniques for small tunnels was carried out in 1964. According to the microclimate and management characters in small tunnels, technical measures such as cultivating mature seedlings in river mud, transplanting in suitable time and covering the tunnel during certain seasons were adopted. These techniques made it possible to harvest small tunnel cultivated cucumber

30 days earlier, prolong the harvest time for 20 days and raise yield up to 75 tons per hectare.

From 1979 to 1981, techniques of viral disease (TMV) prevention and high yield of autumn tomato cultivated in tunnels were studied. By adopting integrated technical measures focusing on exclusion of aphids, the yield of tunnel cultivated tomato rose to over 52.5 t ha^{-1}, with the highest yield 82.5 t ha^{-1}. Viral disease incidence decreased to 1.5%. These techniques were widely popularized in production.

The cultivation experiment of bilayer air charged tunnels was carried out in 1989. Strawberry, tomato, cucumber and other vegetables were cultivated in the similar tunnels. It was applied in production because of its effects in heat preservation, cooling, wind resistance and decreasing the air temperature in tunnel.

The first 2100 m^2 heated greenhouse using factory surplus heat was established in 1980. Temperature in the greenhouse was kept over 17°C in winter, when the lowest temperature in winter might be around -7°C. Cucumber and tomato were successfully cultivated in the heated greenhouse. From 1983 to 1985, tomato was cultivated in the winter season in the greenhouse. Tomato were harvested from December to July and the yield around 150 t ha^{-1}, with the highest yield about 177 t ha^{-1}.

Heated glasshouses and plastic greenhouses were imported from the Netherlands and Israel and located on 15 ha in Nanhui, Pudong, Maqiao, Baoshan and Donghai of Shanghai from 1996 to 1998. They proved successful, especially the Dutch glasshouses more suitable to local conditions. Today, in the glasshouses from the Netherlands, the annual yield of tomato is up to 35 kg m^{-2}, cucumber 40 kg m^{-2} and sweet pepper 19 kg m^{-2}. Cucumbers are exported to Japan. Heated glasshouses based on the Dutch design are made in Shanghai now and are gradually being used by other regions in China.

Vegetable Seedling Production Technique

Seedling raising is the emphasis of vegetable cultivation. Through updating seedling raising facilities and improving seedling raising conditions, the techniques of seedling raising have gradually improved in terms of promoting vegetable production.

In the mid-1960s, raising seedlings under small tunnel conditions was tried and extended. The technical system of using small tunnels to raise seedlings was established. Meanwhile, river mud block, grass pot and other materials were used to raise mature seedlings of fruit vegetables. Electric heating was used to hasten germination. Using these techniques ensured high yield and early harvest.

A study on techniques of raising seedlings under large tunnel conditions was carried out in 1978. Small tunnels, heating wire and other facilities were used in large tunnels to raise seedlings. Seedling pot, film cylinder and other containers gradually replaced river mud block or grass pot.

In 1976, the technique of raising seedlings without soil was tested. Film was selected as the seedling raising pool in a greenhouse, with nutrient solution inside the pool and rice hull as the substrate in the screen. Tomato, cabbage and other vegetable seeds are sown, which grow into small soilless seedlings. The seedlings grow tidily and vigorously with a fine root system and have high survival rates after planting in fields.

During the period 1980-1984 the techniques of industrialized production of seedlings were studied. Temperature control in seedling stage and artificial additional illumination technology were adopted. The P-80 mode temperature-control instrument with four segments was invented to indicate suitable air temperature for raising seedlings of fruit vegetables: 18-14°C are suitable for tomato seedlings, which are short and vigorous at 16-14°C; cucumber prefers 18°C; sweet pepper grows well under 23-21°C. The substrate for raising fruit vegetable seedlings should be vermiculite and fumigated black rice hull. The root system of a soilless seedling is heavier than that of a seedling with soil.

From 1985 to 1988, tests of planting soilless seedlings with substrate succeeded. Grafting technology, cooling and rain-free raising technology in the summer season were studied as well.

In the 1990s, greenhouses equipped with advanced facilities for seedling production and automatic seeding lines were imported from the Netherlands and the USA respectively. Seedling plug techniques were tested and extended effectively.

PROSPECTS FOR DEVELOPMENT OF VEGETABLE PRODUCTION IN SHANGHAI

Primary Mission: Quality Improvement

At present, China is in a new round of reorganizing agricultural structure. Vegetables continue to be the emphasis of this reorganization and the vegetable planting area is still expanding with yield being increased. In some areas, vegetable supply is somehow surplus, while the overall supply and demand of vegetables is generally balanced. Given this situation, Shanghai, considered to have higher production costs than other areas, must make efforts to improve quality in order to obtain good economic results from production and marketing.

Meanwhile, with the opportunities brought by China's entry into the WTO, vegetable export is showing a good growth trend. In 1999, China exported 2.78 million tons of vegetables, earning foreign exchange of USD 1.86 billion. From January to October 2000, vegetable export grew to 2.46 million tons, holding the 4th position after the export of livestock products aquatic products, and grain crops. With the fast growing trend and trade surplus in export, China's vegetable production has good prospects. But the problem at present is that in Shanghai and the whole of China, vegetable quality standard is inferior to international standards. As a member of the WTO, foreign vegetables will enter China's domestic market and domestic vegetable production will face keen competition. Under these circumstances, China must make efforts to improve vegetable quality in order to meet the quality standard of the international market and to maintain healthy development of its own vegetable production.

Increase Input for Scientific Research, Speed the Process of Variety Improvement

For the future, the primary objective of genetic improvement of vegetable variety and new variety breeding is to enhance the quality and disease and insect resistance of *extant* varieties and to develop fine varieties, but not take high yielding as the major consideration in the variety breeding program.

Strengthen International Collaboration and Exchange to Improve Cultivation Techniques

In recent years, international collaboration and exchange has been strengthened further in vegetable production in Shanghai. Experts from the Netherlands, France, Germany, America, Japan, and Israel have been invited to give technical training and consultations to improve the cultivation techniques and management of local producers and people at management level. At the same time many enterprises from different nations have invested in horticultural farms in Shanghai suburbs, producing and selling vegetables to China's domestic and international market. These farms serve as good models of advanced technologies and good management.

Improve Production Environment, Promote Organic Culture

In recent years organic cultivated foods have become the fashion of consumption in the domestic market. It will be the trend for future development as well. By promoting organic culture, which emphasizes providing safe agricultural products while protecting and improving the environment, Shanghai aims to enrich the functions of the metropolis and to enhance its comprehensive competitive power.

Realize Transition in the Mode of Production

As of now, the main mode of vegetable production in the Shanghai area still remains 'family style'. It should gradually be transited to specialization and large-scale production by mostly large horticultural farms, as this would be beneficial to the application and popularization of new technology and new varieties, and would enhance the competitive power of the produce.

SUMMARY

In the past few decades, Shanghai has enjoyed great achievements in vegetable production by continuous adjustment and development. Vegetable supply has risen from shortage in the past to today's balanced demand and supply and entrance into the international market. Vegetable production in Shanghai has played an important role in the agricultural sector, people's life and development of the city. In recent years, vegetable has become a major crop in the Shanghai area. Faced with keen competition from producers at home and abroad, in terms of vegetable production Shanghai must exert more effort in the areas of variety development, improvement, and popularization; seed/seedling industry; improvement of ecological system; promotion of organic culture; quality standardization and enhancement; establishment of a better production base for export; and modernization of vegetable production.

24

CABBAGE DISEASES DURING STORAGE

Fiodor Popov
Belarusian SRI of Plant Protection, Republic of Belarus, 223011, Minsk region, p/o Priluki, BelSRIPP
e-mail: libr@belizr.belzr.belpak.minsk.by

Under storage conditions the most widespread and harmful cabbage diseases are grey mould (*Botrytis cinerea*) and mucous bacteriosis (*Erwinia caratovora*).

Infection of heads of cabbage is promoted by inevitable mechanical damages during treatments, harvesting and transportation, intensified irrigation, especially before harvesting, freezing and annual application of high rates of mineral fertilizers, especially nitrogen. Under increased rates of nitrogen fertilizers the plants have no time to accumulate enough available plant nutrients and are greatly damaged by diseases at storage. Cabbage raised on soils with increased rates of phosphate and potassium fertilizers under moderate nitrogen feeding is more resistant against diseases.

Cabbage losses at storage are explained by the fact that its heads are a favorable medium for phytopathogenic microorganism development. High adaptation and plasticity of pathogens to a feeding substratum and their complicated specific composition create extreme difficulties for breeding cabbage varieties resistant to disease diagnostics of diseases and organization of protective measures. Furthermore, heads of cabbage are colonized by saprophytes and putrefactive microorganisms that speed up rot. In storage grey mould is exhibited as a dirty-gray fluffy bloom on that head surface and causes tissue rotting. With intensive development the decayed leaves become mucous. In many cases the disease turns into a complex rot in which bacterial pathogens participate. Weakened leaf tissues, especially the frozen, mechanically or pest damaged, are the first exposed to grey mould damage.

Mucous bacteriosis is exhibited in two forms: rotting of bract leaves and the whole head and rotting of the cabbage stump. Damaged tissues darken and slime with an unpleasant odour forms on them. Sometimes it appears that the head of cabbage is restratified by healthy and infected leaves.

Grey mould and mucous bacteriosis harmfulness increases with extension and violation of the storage regime. For example, cabbage losses due to grey mould among heads stored in containers under t = 0 + 1°C have reached 22-27% and from mucous bacteriosis 30-35%.

It is known that storage conditions and storage regime exert great influence on the development of diseases. Our observations have shown that the infection rate of heads of cabbage by grey mould and mucous bacteriosis after 4-month storage in storehouses with artificial cooling (t = near 0°C) was about 2-7.3% while that in storehouses with forced ventilation (t = + 1, + 2°C) was 11.7-54.8%. The strongest development of diseases, in particular grey mould, was observed in storehouses without an adjustable regime (t = + 5-6°C), where temperature overfalls caused condensate formation. Storability of late ripening cabbage varieties such as Tyurkiz, Langedeyker, Belaruskaya 85 proved good.

25

FIELD DISEASES OF CUCUMBER AND TOMATO FRUIT

Tamara Zherdetskaya
Blarusian SRI of Plant Protection, Republic of Belarus,
223011, Minsk region, p/o Priluki, BelSRIPP.
e-mail: libr@belizr.belzr.belpak.minsk.by

Scientists at Belarusian SRI of Plant Protection determine through research (1998-2000) that during plant vegetation cucumber and tomato fruits are damaged by infectious diseases.

Late blight. Disease agent—fungus *Phytophthora infestans* de Bary. From surface and inside the tomato fruit the tissue turns brown but remains solid; fungal sporulation does not form.

Grey mould. Disease agent—fungus *Botrytis cinerea* Pers. On tomato fruits small-size, aqueous, light green, round spots with a concentric zone develop. Later the spots merge, their centre turns white and sunken and the edge rises. The tissue begins to rot and is covered with grey sporulation.

Plant-top rot. Disease agent—bacterium *Pseudomonas lycopersici* Burgwitz. At top of green fruit an aqueous spot appears, greyish-green, pale brown and brown, flat or sunken, dry, solid with concentricity.

Ascochytosis. Disease agent—fungus *Corynebacterium michiganense* Jensen. In cucumber fruits the tip is damaged, softened, dries up; on lengthening, many fruits are deformed, can crack, and sometimes are covered with pycnidia.

Bacterial cancer. Disease agent—bacterium *Corynebacterium michiganense* Jensen. With seed infection, tomato fruits exhibit no infection signs.

Sometimes malformed seeds turn dark and lose germination. The secondary local green fruit damage is manifested by whitish spots, by maturation the fruit is not uniformly red and the spots turn yellow in the centre, reminiscent of a 'bird's eye'.

Necrosis of stem medulla. Disease agent—bacterium *Pseudomonas corrygata* Roberts et. Scarlett. In green tomato fruits sometimes 'a net' of whitish small veins forms.

Black bacterial spot disease. Disease agent—bacterium *Xanthomonas campestris* pv. *vesicatoria* Doide. Tomato fruit infection is of the scab type—dark convex points enclosed by a watery border. When the points turn to ulcers, the border may disappear.

Under protected ground conditions in Belarus, during plant vegetation the marketable quality of fruits is worsened by fungal and bacterial diseases: in cucumber—ascochytosis; in tomato—late blight, grey and plant-top rot, bacterial cancer, necrosis of the stem medulla and black bacterial spot disease.

26

THE VARROA MITE AND CROP QUALITY

Ramdane Dris
World Food Ltd.
Meri-Rastilantie 3C
Helsinki, FIN-00980
website: www.food-farming.info or www.world-food.net
e-mail: Dris.uh@Kolumbus.fi

INTRODUCTION

The parasitic mite *Varroa jacobsoni* (Oudemans) infests honey-bee colonies and has become a major obstacle to profitable beekeeping in many parts of the world. Honey-bees derive their protein, vitamins, minerals and some carbohydrates from pollens. Since no single pollen source provides all their nutritional needs, honey-bees must have a number of pollens available to them to remain healthy and to produce the royal jelly required to feed the queen and rear the brood. Insecticides, herbicides and cultivation practices have reduced feral bee populations in many areas. Beekeepers, who feed their colonies high fructose corn syrup, use that syrup to formulate patties that maintain their moisture well because fructose attracts moisture. Fatter bees, with more stored nutrients, winter better and rear more brood the next spring than non-fed bees (Dris, 1999a, b). Growers of horticultural crops rely on honey-bee (*Apis mellifera* L.) colonies to meet their pollination needs. Regardless of colony size, weather factors such as temperature, wind, velocity, solar radiation and relative humidity determine the percentage of a colony's foraging population that will leave the hive to forage at any given time.

In Europe, some areas have been much harder hit and in Britain, for example, soft fruit grower losses have been as high as 90%. This is a serious problem for European farmers, for whom bees are one of the main means of pollinating flowering crops such as oilseed rape, linseed, field beans and fruits. The principal enemy is a tiny mite, only just big enough to be seen with the naked eye, called *Varroa jacobsoni* (Dris, 1999a, b).

LIFE CYCLE OF VARROA

Varroa jacobsoni are small crab-shaped parasites that spend their time either under the cappings in the brood nest or on adult bees. They carry deadly viruses and bury themselves in the sealed cells inside the hive where the larval bees pupate. They feed by forcing their mouthparts through the exoskeleton of the bee and drinking its blood. During the brood rearing season, most of a colony's mites will be buried in the brood. When there is no brood, all the mites attach to adult bees. The bees emerge with damaged legs or wings, or are simply too weak to perform their function in the complex life of the colony. No other problem kills more bee colonies. They are only able to reproduce themselves in the capped brood cells. The mother mite starts to lay eggs about three days after cell capping, one egg every 30 h. The first egg will produce a male and all the successive eggs, females. The time required for a mite to mature is about seven days. So more young mites are produced and, understandably for the mite it makes sense to prefer the drone brood (Dris, 1999a, b).

HOW DID VARROA SPREAD?

The Varroa mite was first introduced in Florida in 1987. Since then, it has spread throughout North America and controlling this parasite has become the number one concern of most beekeepers. By the 1990s it was widespread in continental Europe due to the illegal trading of queen bees to increase stocks. Slower build-ups of mite populations at high altitudes compared to low altitudes in Germany have been reported, suggesting an influence of climate on the population dynamics of the mite. However, results from south-eastern Finland indicate a mite population growth similar to that under central European conditions. Some beekeepers say their bees seem to suffer more, perhaps because the cooler, wetter weather makes the bees more susceptible to the mite, or its associated viruses or both. All infested colonies are expected to collapse without some mite control (Dris, 1999a, b). A rough estimation is that the Varroa population doubles every month in hives in which the bees have capped the brood. Without intervention from the beekeeper, this will lead to breakdown of the colony after one or two years. Varroa has also destroyed millions of wild honey-bees, the ones formerly seen nesting in trees, or sometimes in chimneys. A honey-bee colony can number 50,000 to 80,000 bees. If you have lost 20 colonies in a couple of square miles, that is a lot of pollinators gone, and the impact on the environment will be large.

HOW TO PREVENT DAMAGES CAUSED BY VARROA ?

a) Chemical Treatments

Treatment is available on plastic strips that can be stuck into hives. Beekeepers have also developed a range of traditional treatments, including formic acid (which is still in the development stage, since new methods are still being proposed and tested), marjoram oil and lactic acid. In Europe, thousands of mites shower down in treated hives, looking like little black crabs. But treatments are ineffective over the long term (Dris, 1999a, b). Other colonies become reinfected because drones—the male bees whose main job is to mate with the queen—sometimes move from one hive to another. Losses are about 30% colonies per year.

The fight against Varroa starts after the last crop, which is normally July or August. While the amount of brood is steadily decreasing, the number of mites is continuing to increase; hence the percentage of parasitized brood cells will strongly increase without treatments. These bees will have a reduced lifespan and therefore are not good winter bees (Dris, 1999a, b). Normal winter bees will feed brood and gather nectar up to April of the next year! A high number of mites in August can therefore be catastrophic for overwintering. Studies carried out abroad showed that the majority of hives have a bottom board with a mesh. These bottom boards are provided with a small drawer inserted from the reverse side and after 4-6 days the mites that have fallen in the drawer are counted. One mite per day is roughly equal to 250 mites in the colony if there is brood present and more if there is no brood. If more than one varroa per day falls, a short-time application is done with formic acid. However, studies have shown that the most effective time to treat, from the standpoint of colony health and size, is during the period of new bees being reared for the overwintering population. That generally occurs from mid-August to the end of October. Most commercial operators treat twice a year, just to be sure. It is important to limit the number and lengths of treatments as much as possible to preclude selection for fluvalinate-resistant mites (Dris, 1999a, b).

b) Physical Control

Some physical methods in Varroa control are also practised, such as thermotherapy. When non-sweped bees were treated with heat up to commencement of mortality, which started after 22 min at 45°C, after 16 min at 46°C and just 6 min at 47°C. Efficiency of thermotherapy was nearly 98%.

c) Biological Control

A biological method for Varroa control using wild mites *Chelyetus eruditus* was verified, but these mites did not find conditions suitable in hive space to kill Varroa mites.

Phytotherapy in Varroa control was also verified, using the following plants: garlic (*Allium sativum*), tobacco (*Nicotiana tabacum*), walnut (*Juglans regia*), tomato (*Lycopersicon esculentum*), wormwood (*Artemisia absinthium*), pine (*Pinus sylvestris*), tansey (*Tanacetum vulgare*) and others. An efficiency of 50-80% was attained using these plants when their extracts were applied 2-3 times a week, when a paper smeared with a layer of sticky substances (vaseline, oil) was inserted between the wire mesh and the bottom of the hive.

Chemotherapy of honey-bee colonies is precedent being tested for Varroa control. Both domestic and foreign chemicals are being evaluated, containing such active ingredients as amitraz, nitrobenzene, phenothiazine, thymol, brompropylate, coumaphos, cimizole hydrochloride, formic acid, lactic acid, fluvalinate, acrinathrine, flumethrine and many others with acaricide effect (Dris, 1999a, b).

CONCLUSION AND RECOMMENDATIONS

Infested bees leaving the hive during winter significantly increase the winter mortality rate of Varroa mites. A 50% bee mortality during winter is normal in cold climates and hence mite mortality is greater than in warmer climates with a shorter brood-free mortality. Varroa mites in honey-bee colonies must be controlled if colonies are to survive (Dris, 1999a, b).

Always follow label instructions for handling and applying any chemical in the beehive. Never implement any type of disease or pest treatment program while surplus honey supers are on the hives. The growth rate of Varroa populations observed in south-eastern Finland is comparable to or even exceeds the values observed in Germany. A successful campaign against Varroa should not rely on just one treatment. This may fail in one year due to climatic conditions or incorrect application by the beekeeper, both of which promote resistance to treatments. A combination of two or three different methods, spread over the year, minimizes these possible faults. Also, diagnosis is important in monitoring Varroa populations in the colonies (Dris, 1999a, b).

Many studies have demonstrated that several organic acids already present in small amounts in beehives (lactic, formic, oxalic acid) can be used to control Varroa mites. There is evidence that etheric oils and some common food additives combined with biotechnical management can be effectively used for mite control. However, these ecological alternatives to synthetic acaricides are generally more laborious and show variation in efficacy depending on external factors and colony condition (Dris, 1999a, b). Cultural practices that conserve pollinators ease the stress shortages in colony availability that affects crop size and quality, and ensure the productivity vital to sustaining agroecosystems. The good news for bee lovers is that bumble-bees—the big, striped bees most often seen in gardens—are immune to Varroa, as are more than 100 kinds of solitary bees. But the bad news is that these bees too are becoming less common because agriculture and tidier gardens are destroying their habitat. Anyway, their numbers are much too small to replace the lost millions of honey-bees (Dris, 1999a, b). Thus there is great need for a network of scientists in this field, capable of identifying the research required, to harmonize evaluations of protocols, disseminate research results and make recommendations in a comprehensive form usable as guidelines for beekeepers in Europe.

References

Dris, R. 1999a. Varroa mehiläishoitajan ongelma. Maaseudun Tulevaisuus, 23.9.1999, Nro 109, 83: 2.

Dris, R. 1999b. Vaarallinen varroa. Yliopistolehti. Acta Universitatis Helsingiensis 20: 31.

27

APPROPRIATE HYDROPONIC SYSTEMS FOR GROWING TROPICAL LEAFY VEGETABLES IN GREENHOUSES

W.A.P. Weerakkody[1], N.A.K.G.E. Kumari[1], W.M.S.M. Bandara[1] and R. Dris[2]
[1]University of Peradeniya,
Department of Crop Science, Faculty of Agriculture, Sri Lanka
e-mail: palitha@cropsci.pdn.ac.lk
[2]World Food Ltd., Meri-Rastilantie 3C, Helsinki, FIN-00980
website: www.food-farming.info or www.world-food.net
e-mail: Dris.uh@Kolumbus.fi

INTRODUCTION

In Sri Lanka, which is a tropical country, vegetable cultivation is usually practised under open-field conditions because of low seasonal variations in temperature. However, recent studies have shown that the seasonal fluctuations in vegetable production, post-harvest losses and thus market prices are accountable to unpredictable variations in seasonal rainfall and its secondary influences on crop growth (Weerakkody et al., 1997, 2000). As a consequence of rainfall-related high pest and disease incidence, open field vegetable farmers tend to overuse agrochemicals, leading to high retention of chemical residues, especially in leafy vegetables at the marketable stage (Wijesinghe, 1999). In addition, these vegetables are sometimes contaminated with *E. coli* bacteria and nematodes that cause problems for human health

(Wahundeniya and Kurukulaarachchi, 1999). Under these circumstances, improvements in yield and quality of indigenous leafy vegetables by introducing modern greenhouse technology would be highly remunerative with respect to local and export markets. The majority of protected culture (greenhouse) growers use a single commercial fertilizer called Albert's mixture for growing vegetables in hydroponics culture. Compared with alternative hydroponics fertilizers, it has proven to be better for popular tropical leafy vegetables in trough culture (open-type non-aggregate systems) (Wijesinghe, 1999). However, its performance in other commonly used hydroponic systems for the growth and leaf yield of leafy vegetables together with fertilizer-use efficiency have to be examined.

MATERIAL AND METHODS

Research was conducted on kankung (*Ipomea aquatica*) and mukunuwenna (*Alternanthera sessilis*) at the Horticulture Research and Development Institute at Gannoruwa in Kandy District. The location, elevation, temperature and mean annual rainfall of the site are 7° 17′N, 80° 28′E, 488 a.m.s.l., 24.4°C and 2088 mm respectively. The experimental design was a completely randomized block design (CRD) with 4 treatments and 3 replications. Three soilless cultures (hydroponic systems)—NFT, coir bag and deep troughs—were compared with soil culture.

All the treatments (cultivation patterns) were arranged in a barrel-shaped tunnel greenhouse with elevated roof vent (top vent), and galvanized iron frame. It was covered with 200 μm UV-inhibitor transparent polythene. The walls were covered with 100% insectproof nets, to a height of 240 cm from ground level. The height, width and length of the tunnel were 6 × 9 × 30 m. Two high-mounted exhaust fans were used for forced-air ventilation at the two ends. As an additional temperature control device, a mist system was arranged in two parallel lines with hanging-type microsprinklers along the length of the greenhouse.

The nutrient solution used to fertigate all the treatments was prepared with Albert's mixture (Unipower, 1999) and kept in a 500 L catchment tank; pH was maintained between 6.5 and 7.0. The electrical conductivity (EC) was adjusted to 1200 mhos and pH and EC of the nutrient solution of the catchment tank were determined daily and adjusted at frequent intervals. Grow bags and troughs were operated as open-type hydroponic systems while the NFT was operated as a closed-type (recirculating) without recovery of residual nutrients.

The circulating tubes of the NFT system were 8 cm diameter and 9 m long polyvinyl chloride (PVC) pipes placed on a 1 m high wooden stand at a slope of 100:75. Plants grown in coir dust-filled plastic net pots were fixed into 2.5 cm diameter holes in the pipe at 20 cm distances. The arrangement of coir bags and troughs was similar to that described by Wijesinghe (1999). Drip tubes were used to provide nutrients continuously to grow bag culture while the nutrient solutions in troughs were replaced fortnightly. For soil culture, raised beds were prepared on the greenhouse floor after pulverizing the soil. The same nutrient solution was used to supplement the nutrients for the soil culture.

Semihardwood cuttings were used as planting materials of kankung and mukunuwenna. Crops were managed according to the recommendations of the Department of Agriculture (Technoguide, 1990). Plant leaf area, cumulative plant height (CPH) and number of leaves were the parameters used to evaluate growth performance during the vegetative period.

At the end of the 4th harvest root fresh weight and dry weight were determined. Fresh and dry weights of the yield were also measured. Dry weight was obtained by drying the harvested leaves after recording fresh weight in an oven at 105°C temperature when they reached constant weight. Data analysis was done using computer software SAS.

RESULTS AND DISCUSSION

Kankung

The effect of cultivation pattern on cumulative plant height, mean leaf area and number of leaves per plant was statistically significant 9-15 days after planting (DAP). In general, late vegetative growth and leaf yields of hydroponically grown kankung were greater than soil culture. Significantly higher plant height was observed in NFT than in the other two cultures. The highest mean leaf area was found in the NFT and grow bag culture. The highest mean leaf number was found in the grow bags at final harvesting. In terms of combined mean leaf area and number of leaves per plant, NFT gave the greatest leaf area per plant (during 9-15 DAP). Thus the yields of NFT and grow bag plants were greater than in the other cultures (Fig. 27.1).

However, the root growth of kankung appeared to differ slightly from shoot growth. The highest root fresh weight was found in plants grown in troughs and NFT at the final harvesting stage (45.85 and 38.93

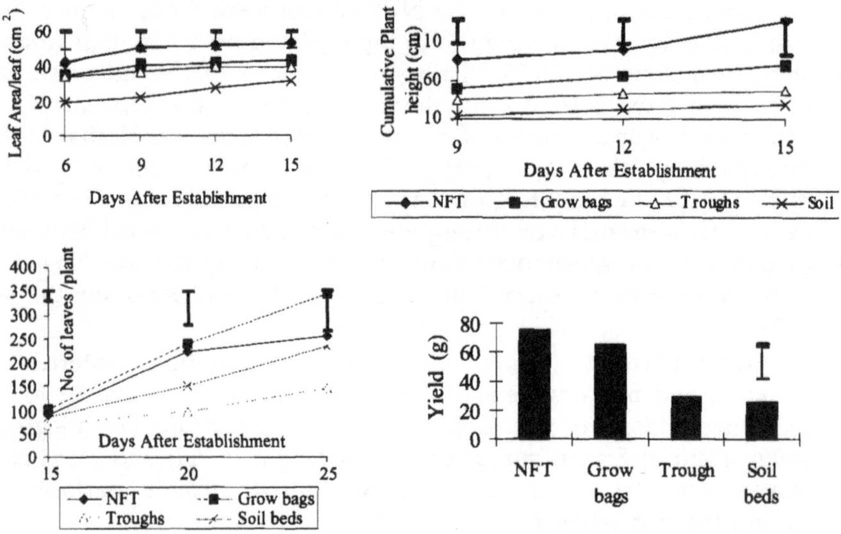

Fig. 27.1 Leaf area (*top left*), yield (*top right*), leaf number (*bottom left*) and plant height (*bottom right*) of Kankung in hydroponics and soil culture. (LSD values are indicated as vertical bars)

g plant^{-1} respectively). In contrast with the other growth parameters, the dry matter content of the leaf yield was higher in soil culture than in the hydroponics.

Usually kankung grows in flooded damp places in the wet zone, and shallow water and moist places in the dry zone. This means it is an aquatic or semiaquatic plant form (Rice et al., 1993). Normally such places are deficient in nitrogen sources due to the activity of anaerobic denitrification bacteria. Hence significantly lower leaf yield of the soil-grown plants could be a result of nutrient toxicity, such as nitrogen, that occurred under application of the plant nutrients at the same strength and dosage as used in the other treatments (grow bag culture). Comparatively dark green leaves observed in the soil plots supports this justification (high N content). Similar observations reported by Bould and colleagues (1983) lead to the speculation that the major reason for this difference was N or Mg toxicity. On the other hand, the relative decline in shoot growth in troughs could be related to the imbalance in root-to-shoot ratio.

Mukunuwenna

The cultivation pattern did not affect cumulative plant height and leaf area significantly. Number of leaves per plant, however, showed a significant treatment effect, beginning from 20 DAP. Generally, more leaves could be observed in NFT and grow bag cultures, with the latter dominant at the final stage. Higher mean leaf number was obtained from mukunuwenna plants grown in NFT and grow bags than in the other treatments. The highest number was found in grow bags at the final stage. Hence grow bags (45.41g) and NFT system (40.32 g) gave the highest leaf yield (Fig. 27.2).

Wahundeniya and Kurukulaarachchi (1999) indicated that mukunuwenna prefers well-moisten soil with good drainage. The well-aerated conditions with adequate moisture availability in grow bags and NFT appeared to be more conducive for greater leaf production of mukunuwenna. Furthermore, the above properties in grow bags and NFT appeared to be favourable for a balanced root-to-shoot ratio and thus for a higher yield of mukunuwenna. That could be the reason why mukunuwenna plants grown in grow bags performed better than in the liquid (non-aggregate) hydroponic system (trough culture).

Similar to kankung, mukunuwenna showed a significantly higher dry matter content in plants grown in soil than in hydroponics.

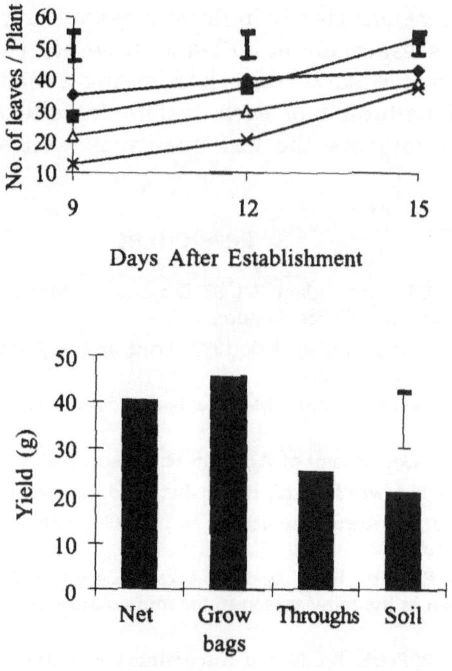

Fig. 27.2 Leaf growth (*top*) and yield (*bottom*) of mukunuwenna in hydroponics and soil culture (Least significant differences [LSD] values are indicated as vertical bars)

Table 27.1 Time of harvesting and harvesting interval of leafy vegetables grown in different cultivation patterns

Crop	Soil Cultivation		Hydroponic Systems	
	1st Harvest (DAP)	HI Days	1st Harvest (DAP)	HI Days
Mukunuwenna	28	21-28	28	15
Kankung	30	20	20	10

Relatively rapid water accumulation in leaf tissues could be the reason for the bulky nature of hydroponically grown plants. When considering the inverse relationship between electrical conductivity (EC) of the nutrient solution and water absorption by plants (Schwarz, 1995), little increments in the EC of the solution could be suggested as responsible for increase in the dry matter content of hydroponically grown plants.

Early maturity in hydroponically grown plants provides additional advantages in terms of early harvesting and short harvesting intervals (HI) (Table 27.1) compared with the usual harvesting schedule specified by the Department of Agriculture for open-field cultivation of both crops (Technoguide, 1990).

Hence, the results clearly indicate that kankung and mukunuwenna in hydroponics, especially in NFT and grow bag cultures, exhibit faster growth and greater marketable yield, giving a satisfactory turnover for the grower. Fine-tuning of each system for efficient use of fertilizer would further improve the leaf quality as well as the economics of production.

References

Bould, C., Hewitt, E.J. and Needam, P. 1983. Diagnosis of Mineral Disorders in Plants. Her Majestry's Stationary Office, London.

Rice, R.P.,Rice,L.W. and Tindall,H.D. 1993. Fruit and Vegetable Production in Warm Climates, Macmilian, London.

Schwarz, M. 1995. Soilless Culture Management. Advance Series in Agriculture Science. Springer, NY.

Technoguide. 1990. Department of Agriculture, Peradeniya.

Unipower. 1999. Unipower Pvt.Ltd, Rajamalwtte Rd. Battaramulla, Sri Lanka.

Wahundeniya, K.B. and Kurukulaarachchi, N. 1999. Cultivation of Leafy Vegetables. Dept. Agic., Peradeniya.

Weerakkody, W.A.P., Peiris, B.C.N. and Jayasekara, S.A.B.J. 1997. Yield and fruit quality of tomato grown in the rainy season in the mid-country in Sri Lanka. Trop. Agric. Res. 9: 158-168.

Weerakkody, W.A.P., Peiris, B.C.N. and Gunaratne, L.H.P. 2000. Commercial gardening of vegetables in Sri Lanka: Present status and environmental constraints. Sri Lankan J. Agric. Sci. 37: 135-147.

Wijesinghe, W.A.M.M.K. (1999). Effect of different fertilizer mixtures on the growth and yield of gotukola and kankung in hydroponic culture. B.Sc. Diss., Fac. Agric. Univ. Peradeniya, Peradeniya.

28

CURRENT SITUATION OF GREENHOUSE ENGINEERING IN SHANGHAI

Wenbin Zhang
China Agricultural University,
727 Jiaozhou RD, Shanghai, P.R. China
e-mail: zhwbin@public3.sta.net.cn

INTRODUCTION

To raise the level of Shanghai agricultural modernization and to reduce the differences between our country and advanced agriculture abroad, greenhouses from Holland and Israel were introduced in Shanghai in 1996. One 600 m^2 greenhouse from Holland is located in Donghai vegetable model base (Fan, 1998; Zhu, 1998) and another in Sunqiao modern agricultural development zone of Pudong new district (Sunqiao). The 900 m^2 plastic greenhouses from Israel are situated in Maqiao horticultural farm of Minghan district (Maqiao), Luodian horticultural farm of Daoshan district and the Xinchang Vegetable Horticulture Co. Ltd. of Nanhui county (Pang and Cai, 1998; Cai Xiangyuan et al., 1998). Three years of protected culture practice have yielded failures as well as successes. The knowledge gained from both is summarized below.

FUNCTIONING OF THE MODERN 'FOUNDER' GREENHOUSES

The 'founder' (base) modern greenhouses have functioned normally since 1996, with each year showing improvement over the last.

1. Yield increased year after year. The total yield of the four bases (1200 m^2) (note: Nanhui is not included because it belongs to the Shanghai Agriculture and Forestry Bureau) was 1483.47 t from 1996-1997, in the second year 1633.17 t and in the third year 1995.81 t. The annual increases were 10% and 22.22% respectively.
2. Sales income increased in successive years. The total sales income of the four bases was 6.09 million yuan 1996-1997, 7.57 million yuan in the second year and 9.26 million yuan in the third year. The annual increments were 25.36% and 22.33% respectively (Pang and Cai, 1998).
3. Operational costs reduced gradually. The operational costs of the four bases was 6.28 million yuan for 1996-1997, 7.88 million yuan in the second year and 7.33 million yuan in the third year. Operational costs were less by 6.97% in the third year than in the second.
4. Profit increases remarkably. The Sunqiao and Donghai bases showed a surplus while the other two bases sustained some loss in the first two years. In the third year, however, all four bases showed a surplus. As of 9 June 2000, the total yields and sales income of the four bases were 1742.91 t and 1.04 million yuan—an increase of 1.82% and 25.6% respectively over the previous year. Thus the four bases with foreign advanced greenhouses show great potentiality for yield and enhanced income.

The 3-year practice in Sunqiao exemplifies the results obtained with foreign modern greenhouse culture in Shanghai. Under the supervision of foreign experts, the total yield and sales income in Sunqiao in the first year were 657.46 t and 3.07 million yuan, with mixed average yield 21.91 kg per square metre and average sales income 68.1 thousand yuan per mu. In the second year the yield was 554.54 t and sales income 3521.8 thousand yuan. In the third year the total yield and sales income were 762.53 t and 4501.7 thousand yuan respectively, with average yield 25.42 kg per square metre and average sales income 100 thousand yuan per *mu*. After deducting depreciation charges of 1 million yuan, Sunqiao had a surplus of more than 400 thousand yuan.

Sunqiao also enjoyed income from various businesses, e.g. sale of tickets to various functions, about 2 million yuan; supermarket income 1.75 million yuan; restaurant income 0.67 million yuan; parking facility income 0.11 million yuan; sales of other vegetables 5.34 million yuan etc.

Thus in these three years Sunqiao not only realized a remarkable increase in greenhouse yields and economic benefits, but also exhibited the charm and comprehensive benefits of a modern urban agricultural society (Cai et al., 1998; Fan, 1998).

The Sunqiao modern agricultural development zone has become a part of Shanghai urban tourism and a model zone for modern agriculture. It is estimated that in 1999 more than 350 thousand people visited Sunqiao and nearly 50 thousand high school students became acquainted with the technologies pertinent to the facilities for raising vegetables and flowers.

The Sunqiao modern agricultural development zone has also become a research centre for understanding foreign modern greenhouse culture A series of research programmes have been designed in collaboration with various industries, institutes and universities.

Introduction of the foreign modern greenhouse has also led to the development of related industries, such as greenhouse covering materials, plant varieties suitable for protected culture, irrigation equipment and its technology etc. Obviously the introduction, understanding and assimilation of foreign modern greenhouses has improved horticultural technologies and concomitantly provided a fresh spurt to Shanghai's economic development.

UNDERSTANDING FOREIGN MODERN GREENHOUSES

1. Correct introduction, understanding and assimilation of foreign modern greenhouse culture. The modern greenhouse and its culture technology differs drastically from our traditional approach in many aspects, e.g. equipment and facilities, plant varieties, environmental control, fertigation, insect and disease prevention, gathering, processing and packing of produce. There are many things to be learned and we must study each positively, earnestly and modestly.
2. Realize fully the characteristics of modern greenhouse production. This means taking full advantage of advanced facilities and technologies, creating thereby the best environmental conditions for plants such as temperature, light, fertilizer, water, humidity and CO_2. It further involves raising soil production rate, resource utilization, labour rate and product commodities in order to maximize economic, ecological and social benefits. Great investments must be made to obtain good quality and high yield, and for this we must change our production ideas (Fan, 1998; Zhu, 1998).
3. Low yield is the primary factor restricting benefits. Of the four bases, the benefits realized by Sunqiao and Donghai are better

than those of Maqiao and Luodian. The primary reason for this is that the yields obtained in Maqiao and Luodian were half those realized in Sunqiao and Donghai.

4. Understand modern facility cultivation technology. This is essential to developing the potentiality of advanced facilities and superior varieties in modern horticulture.
5. Extend sales channels and raise business benefits. As the proverb says: 'To grow well is not better than to sell well.' Donghai's yield was the same as Sunquiao's, but Sunquiao's sales income was 1.2 million yuan more than Dhonghai's. Thus much must be done in the area of increasing business benefits.
6. Improve business management, reduce operational costs and increase investment-to-production rate. In 1998 and 1999, the four bases differed in these respects. The lowest operational costs of the 300 m^2 greenhouse was 0.97 million yuan, the highest 3.08 million; the lowest sales income was 0.97 million yuan, the highest 4.50 million; and the lowest rate of investment to production was 1:1, with the highest 1:1.46. Thus we must strengthen management and reduce costs in order to increase economic returns (Pang and Cai, 1998; Zhu, 1998).

PROPOSAL FOR GREENHOUSE ENGINEERING

Greenhouse engineering in Shanghai has revealed its capabilities and concomitantly its value in terms of model benefit, scientific benefit, talent benefit and economic benefit. We are convinced that through hard work for some years to come, we can attain a new level of modern construction in Shanghai. To reach this goal the following suggestions are offered:

1) further improve management, yield and benefits of the modern greenhouse and thereby expand further the potentiality of advanced facilities and technologies;
2) strive mightily for home modernization and industrialization of the greenhouse;
3) strengthen organization and harmonization.

References

Cai Xiangyuan, Zhang Wenbin and Li Hanqing. 1998. The present situation, prospects and proposals for ensuring correct development of the vegetable greenhouse in Shanghai. Trans. Chinese Soc. Agric. Engineering, 14 (supplement): 6-11.

Fan Deguan. 1998. Actualities and futurities of development of Shanghai city-type modern agriculture. Acta Agric. Shanghai, 14 (supplment): 1-10.

Pang Lian and Cai Xiangyuan. 1998. Present situation and development prospects of vegetable greenhouses in Shanghai suburbs. Acta Agric. Shanghai, 14 (supplement): 11-20.

Zhu Songhua. 1998. Theory and Practice of Agricultural Modernization, 12 (1): 413-424. Shanghai Finance and Economic Press.

29

RECUPERATION OF 'BIOCO$_2$' TO INCREASE GREENHOUSE PRODUCTION

Laurent Urban, Loïc Barthélémy, Philippe Bearez, Pierre Pyrrha and André Jaffrin
Unité de Recherche Intégrée en Horticulture
Institut National de la Recherche Agronomique
Route des Colles, Sophia-Antipolis 06410 Biot France

INTRODUCTION

In developed countries, greenhouse horticulture has come to a standstill. Yields are close to their maximum thanks to the widespread use of performing techniques such as supplemental lighting and soilless culture. Meanwhile prices decrease as the result of growing competition from newcomers from Eastern Africa and South America, threatening the future of modern greenhouse production (Urban, 1997a). The prospect of injecting CO_2 all year round all day long could change this by boosting yield and quality of production (Urban, 1995). New sources of CO_2 are under scrutinity both in the Netherlands and France. INRA is paying special attention to biogas as a potential source of cheap CO_2 (Urban, 1995; Bentounes, 1995) because organic wastes from urban activities or the food industry are abundant all over the country. When burned, biogas, a mixture of methane, CO_2 and a few secondary gases, yields mainly water vapour and CO_2 which are released into the atmosphere where the latter contributes (to a lesser extent than methane) to the global change problem.

The benefits of elevated CO_2 are well known and there is a large body of evidence that while anticipated yield increases from extra CO_2 in the terrestrial atmosphere might not eventuate (Enoch, 1990; Bowes, 1993; Murray, 1997), most horticultural crops can take advantage of it without experiencing any of the limiting factors occurring in open-field conditions such as mineral and water availability (Bowes, 1993), and sink activity (Arp, 1991; Bowes, 1991). Harmful UV-B radiations are significantly reduced by glass or plastic material coverings.

In the case of rose crops, no acclimation has been observed to long-term enrichment in CO_2 either by Balazsovits et al. (1989) or Jiao et al. (1991) or Beeson and Graham (1991). Our own observations showed that the number of bud breaks and marketable yield were increased by 23% and 19% over a four-month period of CO_2 enrichment (Urban, 1994). Stem length was improved by 20% (Urban and Langelez, 1992).

But a few questions still remain unsolved. To make use of even cheap CO_2 economically viable injection must be performed in near-to-tight greenhouses. Again, greenhouses must be tight if they are to be used to fix CO_2 for depollution purposes. Unfortunately, the temperature inside greenhouses tends to be much higher than outside during daytime because of the greenhouse effect. High temperatures are known to accelerate photoinhibition or to induce photoinhibition in plants exposed to high photon flux densities (Bongi and Long, 1987).

This paper presents some results of a more general programme aiming at assessing the possibility of using almost closed rose greenhouses to fix CO_2 even in hot conditions, such as those prevailing in summer in southern France.

MATERIAL AND METHODS

Trials were conducted with 'First Red' rose plants grown in two greenhouses: control and one with CO_2. Cuttings were rooted in rockwool cubes on 28 November 1999 and planted in perlite containers at a density of 7 plants per m^2 on 20 January 1999. Recording of data commenced from 20 April 1999.

Plants were irrigated and fertilized according to Urban (1997b) to preclude any water and salinity stress. The pH was set at 5.6 and electrical conductivity at 1.4 mS cm^{-1}.

CO_2 was injected from a liquid gas tank (AGA) so as to maintain 700 vpm throughout daytime under the control of a Siemens gas analyser.

Windows had a fixed aperture of 5% and temperatures were allowed to rise unlimited. Climatic parameters were continuously recorded by an

INRA-made system (SYSPIL). We checked that average temperatures remained in close range in both greenhouses.

Leaves used for fluorescence and gas exchange measurements were chosen according to the selection criteria described by Urban and Langelez (1992).

Ambient CO_2 concentration, photosynthetically active photon flux (PAR), leaf temperature (θ_{leaf}), net CO_2 uptake (A) and leaf diffusive conductance (g_s) were monitored with a portable, open system, infra-red analyser LCA-2 (ADC Ltd., England) and a leaf chamber (Parkinson, 1983) on 22 June 1999. About 30 s were required for equilibration of CO_2 depletion in the differential mode. Calculations were performed according to the models of Von Caemmerer and Farquhar (1981).

Fluorescence parameters were measured on 28 June 1999 on intact, attached leaves by a portable modulated fluorometer (FMS, Hansatech Ltd., U.K.) after dark adaptation for at least one hour, provided by dark clips. The F_v/F_m parameter (Krause, 1988) was calculated as an indicator of photoinhibition. The daily variations in PAR and temperatures were similar to those of 22 July (data not presented).

Results are expressed as means ± SE. The t-test was used to determine significance of differences ($P < 5\%$).

RESULTS AND DISCUSSION

There were no significant differences in PAR and θ_{leaf} between the CO_2 greenhouse and control (Figs. 29.1 and 29.2). Leaf temperatures rose up to 38°C.

It is obvious from Figure 29.3 that stomata of 'First Red' rose plants do not close in response to elevated CO_2, confirming earlier observations by Urban (1992) on 'Sonia'. Stomata of most species close as a short-term response to elevated CO_2 (see Murray, 1997, for a review of the topic). Although unusual, our observation is not an isolated one. Bunce (1992) observed that stomata of *Malus sylvestris, Quercus prinus* and *Quercus robur* were unaffected by elevated CO_2, as did Conroy et al. (1988) on *Pinus radiata.* Hibbs et al. (1995) even observed an increase in stomatal conductance in red alder leaves at elevated CO_2. There are no explanations for such discrepancies.

Net CO_2 assimilation by leaves was significantly increased at 9h45, 12h00, 14h00 and 16h00 in the elevated CO_2 treatment. The increase reached 140% at noon. We observed in an earlier trial that 900 vpm results in an increase of near 200% in leaf net photosynthesis and of about 57% in dry matter production in 'Sonia' rose plants grown soilless

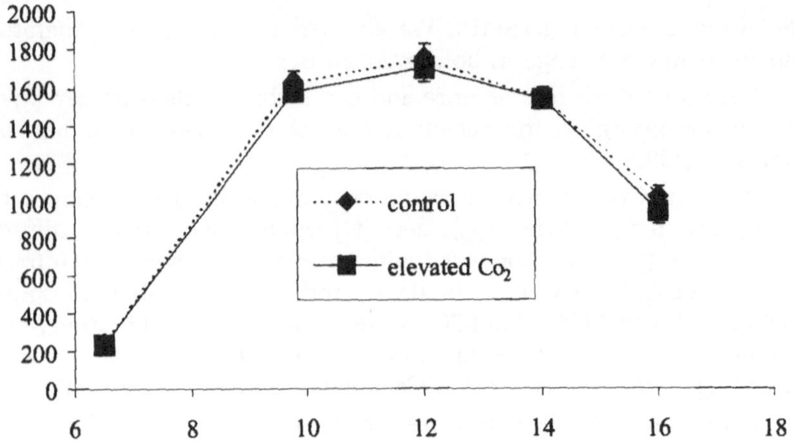

Fig. 29.1 Daily variation of PPF. Vertical bars represent standard errors.

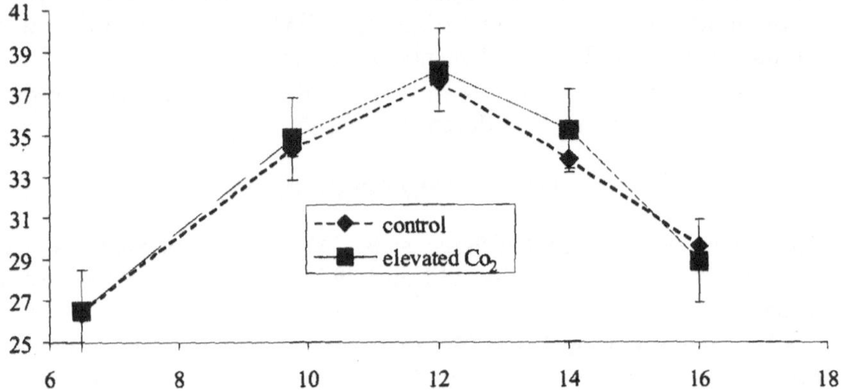

Fig. 29.2 Daily variation of leaf temperature (°C). Vertical bars represent standard errors.

with truly non-limiting water and mineral supply (Urban, 1994). All these figures appear quite high compared to data from the literature (see Murray, 1997 for a review). According to the model of Bunce (1998), A should increase by only 110% in response to 700 vpm CO_2/35°C. Bunce's model makes the usual assumptions about the specificity of Rubisco and of a constant internal on external CO_2 concentration ratio. These assumptions may be questioned. We also feel that the hypothesis of an inhibitory effect of elevated CO_2 on dark respiration should not be overlooked (Idso and Kimball, 1993a, b; Griffin et al., 1996).

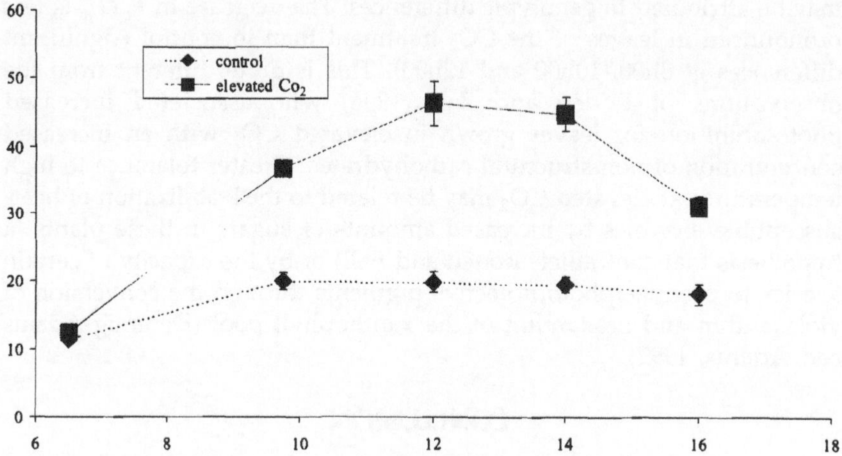

Fig. 29.3 Daily variation of leaf diffusive conductance. Vertical bars represent standard errors.

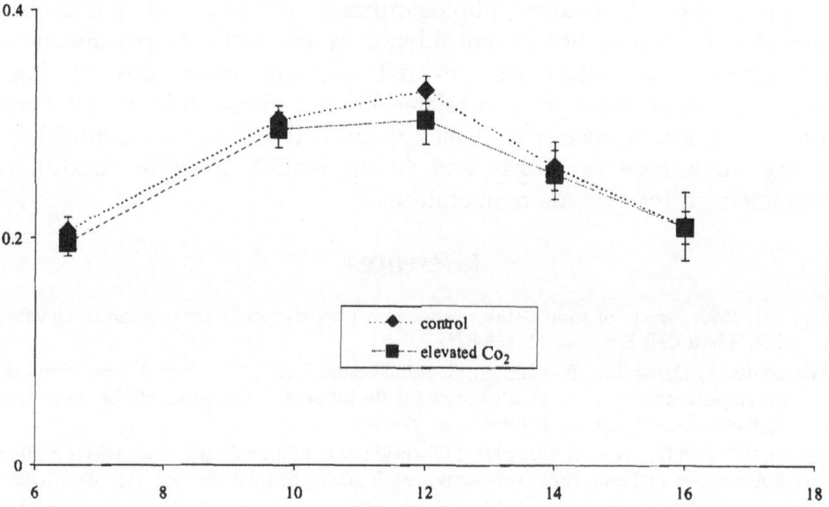

Fig. 29.4 Daily variation of net CO_2 fixation rate. Vertical bars represent standard errors.

Morning figures of F_v/F_m are very high (Fig. 29.4), but not unrealistic. Maury (1997) found F_v/F_m figures as high as 0.857 in well-watered sunflower plants grown in greenhouses. The decrease observed afterwards appears less pronounced than in the two well-watered eucalyptus species studied by Roden and Ball (1996). Such differences may be attributed to genotypic differences. The decrease in F_v/F_m is less pronounced in leaves of the CO_2 treatment than in control (significant differences at 8h00, 10h00 and 12h00). This is quite different from the observations of Roden and Ball (1996) who associated increased photoinhibition for leaves grown in elevated CO_2 with an increased concentration of non-structural carbohydrates. Greater tolerance to high temperatures at elevated CO_2 may be related to the stabilization of heat-susceptible enzymes by increased amounts of sugars in these plants (a hypothesis that contradicts Roden and Ball) or by the capacity of certain species to produce photoprotective pigments through the conversion of violaxanthin and zeaxanthin of the xanthophyll pool (Demmig-Adams and Adams, 1992).

CONCLUSION

Photosynthesis in rose plants strongly benefits from CO_2 enrichment. There is a large body of evidence that elevated CO_2 can help roses to overcome some of the negative effects associated with high temperatures. Of course, photosynthesis, although of paramount importance, should not be considered as the ultimate parameter of production. The effect of elevated CO_2 in conditions of high temperatures on vase life and related criteria needs to be investigated too. If satisfactory, roses may well appear as the perfect crop-model for taking advantage of biogas and fixing bioCO_2 even in conditions characterized by extreme temperatures.

References

Arp, W.J. 1991. Effects of source-sink relations on photosynthetic acclimation to elevated CO_2. Plant Cell Environ. 14: 869-875.

Balazsovits, J., Dansereau, B., Gagnon, S. and Beeson, R.C. 1989. Effets à long terme de l'enrichissement carboné et d'un appoint de lumière sur la productivité de la rose cultivée en serre. Can. J. Plant Sci. 69: 979-989.

Beeson, R.C. and Graham, M.E.D. 1991. CO_2 enrichment of greenhouse roses affects neither Rubisco nor carbonic anhydrase activities. J. Amer. Soc. Hortic. Sci. 116: 1040-1045.

Bentounes, N. 1995. Valorisation du biogaz par enrichissement en CO_2 des serres horticoles. PHM Rev. Hortic. 358: 13-17.

Bongi, G. and Long, P.S. 1987. Light-dependent damage to photosynthesis in olive leaves during chilling and high temperature stress. Plant Cell Environ. 10: 241-249.

Bowes, G. 1991. Growth at elevated CO_2–photosynthetic responses mediated through Rubisco. Plant Cell Environ. 14: 795-806.

Bowes, G. 1993. Facing the inevitable: plants and increasing CO_2. Ann. Rev. Plant Physiol. 44: 309-332.

Bunce, J.A. 1992. Stomatal conductance, photosynthesis and respiration of temperate deciduous tree seedlings grown outdoors at an elevated concentration of carbon dioxide. Plant Cell Environ. 15: 541-549.

Bunce, J.A. 1998. The temperature dependence of the stimulation of photosynthesis by elevated carbon dioxide in wheat and barley. J. Exp. Bot. 326: 1555-1561.

Conroy, J.P., Milham, P.J., Bevege, D.I. and Barlow, E.W.R. 1988. The influence of CO_2 enrichment, phosphorus deficiency and water stress on the growth, conductance and water use of *Pinus radiata* D. Don. Plant Cell Environ. 11: 91-98.

Curtis, P.S. 1996. A meta-analysis of leaf gas exchange and nitrogen in trees grown under elevated carbon dioxide. Plant Cell Environ. 19: 127-137.

Demmig-Adams, B. and Adams, W.W. 1992. Photoprotection and other responses of plants to high light stress. Ann. Rev. Plant Physiol. Mol. Biol. 43: 599-626.

Enoch, H.Z. 1990. Crop responses to aerial carbon dioxide. Acta Hortic. 268: 17-32.

Griffin, K.L., Ball, J.T. and Strain, B.R. 1996. Direct and indirect effects of elevated CO_2 on whole-shoot respiration in ponderosa pine. Tree Physiol. 16: 33-41.

Hibbs, D.E., Chan, S.S., Castellano, M. and Niu, C.H. 1995. Response of red alder seedlings to CO_2 enrichment and water stress. New Phytol. 129: 569-577.

Idso, S.B. and Kimball, B.A. 1993a. Effects of atmospheric CO_2 enrichment on net photosynthesis and dark respiration rates of three Australian tree species. J. Plant Physiol. 141: 166-171.

Idso, S.B. and Kimball, B.A. 1993b. Effects of atmospheric CO_2 enrichment on photosynthesis, respiration and growth of sour orange trees. Plant Physiol. 99: 341-343.

Jiao, J., Tsujita, M.J. and Grodzinski, B. 1991. Influence of radiation and CO_2 enrichment on whole plant net CO_2 exchange in roses. Can. J. Plant Sci. 71: 245-252.

Krause, G.H. 1988. Photoinhibition of photosynthesis. An evaluation of damaging and protective mechanisms. Physiol. Plant. 74: 566-574.

Maury, P. 1997. Adaptation à la sécheresse et photosynthèse chez le tournesol (*Helianthus annuus* L.). Ph.D. thesis, Institut National Polytechnique de Toulouse, 137 pp.

Murray, D.R. 1997. Carbon dioxide and plant responses. In: P. S. Nutman (ed.). Research Studies in Botany and Related Applied Fields. Research Studies Press Ltd., Taunton, Somerset, England, 275 pp.

Parkinson, K.J. 1983. Porometry. S.E.B. Symposium on Instrumentation for Environmental Physiology. Cambridge University Press, pp. 171-191.

Roden, J.S. and Ball, M.C. 1996. The effect of elevated CO_2 on growth and photosynthesis of two eucalyptus species exposed to high temperatures and water deficits. Plant Physiol. 111: 909-919.

Urban, L. 1992. Effect of high pressure mist and daytime continuous CO_2 supplementation on water status and quality of 'Sonia' rose plants grown on rockwool. Adv. Hortic. Sci. 6: 77-81.

Urban, L. 1994. Effect of high pressure mist and daytime continuous CO_2 enrichment on leaf diffusive conductance, CO_2 fixation and production of *Rosa hybrida* plants grown on rockwool. Acta Hortic. 361: 317-324.

Urban, L. 1995. Le biogaz, un ballon d'oxygène pour l'horticulture ? PHM Rev. Hortic. 358: 5-12.

Urban, L. 1997a. Introduction à la production sous serre. Tome 1. La gestion du climat. Lavoisier (ed.). Paris-London-New York. ISBN 2-7430-0205-0, 306 pp.

Urban, L. 1997b. Introduction à la production sous serre. Tome 2. L'irrigation fertilisante en culture hors sol. Lavoisier (ed.). Paris-London-New York. ISBN 2-7430-0206-9, 210 pp.

Urban, L. and Langelez, I. 1992. Effect of high-pressure mist on leaf water potential, leaf diffusive conductance, CO_2 fixation and production of cultivar 'Sonia' rose plants grown on rockwool. Sci. Hortic. 50: 229-244.

Von Caemmerer, S. and Farquhar, G.D. 1981. Some relationships between the biochemistry of photosynthesis and the gas exchange of leaves. Planta 153: 376-387.

30

MECHANISTIC MODELLING PREDICTIONS OF SUGAR-CANE YIELDS IN SEVERAL LATIN AMERICAN COUNTRIES

Angel Utset[1], José López[2], Manuel Alvarez[3], María Elena Ruiz[1] and Alicia del Valle[1]
[1]Agrophysics Research Unit, Agricultural University of Havana (UNAH), Apdo. 18, San José de las Lajas, La Habana 32700, Cuba
[2]Institute of Geography, University of Mexico (UNAM)
[3]University of Los Llanos "Ezequiel Zamora" (UNELLEZ)
e-mail: utsets@yahoo.com

INTRODUCTION

Sugar-cane is cropped in tropical and subtropical areas of the world, where precipitation behaviour is seasonally distributed. Hence, irrigation supply is a largely recognized necessity for obtaining higher yields, particularly in the first sugar-cane development stages (Reynoso, 1862; Thompson, 1977; Fonseca, 1984; Wiedenfeld, 2000). In many sugar-cane-cropping places of the world, irrigation is expensive and water scarce. Therefore, farmers must find a profitable relationship between irrigation supply and sugar-cane yields. Presently available crop simulation models offer agricultural decision-makers an excellent tool for selecting the water management solution that would achieve the most reliable yields under variable soil or weather conditions (Burke et al., 1999; Singh et al., 1999). Nevertheless, modelling predictions are sometimes inaccurate (Landau et al., 1998; Richter, 1999), which diminishes model confidence for decision-making and points out the need for model validations under wide ranks of modelling inputs.

Several modelling approaches are available regarding water availability effects on crop yields. It has been shown that modelling predictive accuracy depends on model assumptions concerning root water uptake (Jamieson et al., 1998). Mechanistic models, i.e., those based on the physical laws of the soil-water-plant-atmosphere continuum, are able to simulate capillary rising, fast and low drainage and several other features that yield more accurate simulation results than non-mechanistic models (Gabrielle et al., 1995; Maraux and Lafolie, 1998; Singh et al., 1999). In particular it has been shown that sugar-cane roots are able to extract water from deeper depths than the root-active zone under rainfed conditions (Rivero, 1985) and hence the crop irrigation requirements could be overestimated by using models that do not take into account capillary rising. Most of the modern mechanistic models were developed for temperate-climate crops. However, some attempts have recently been made to provide the sugar-cane parameters needed in mechanistic modelling estimations of sugar-cane yields (Keating et al., 1999; Ruiz et al., 1999). The performance of such sugar-cane mechanistic models has yet to be tested under different conditions, however. On the other hand, combining the model with a Geographic Information Systems (GIS) allows not only regionalization of the model outputs, but also scaling up of model inputs (Rosenthal et al., 1998; Richter, 1999). Moreover, since mechanistic models are very sensitive to the spatial variability of soil hydraulic properties, i.e., soil-water retention curve and saturated hydraulic conductivity (Leenhardt et al., 1995; Connolly, 1998), this variability could be characterized before modelling using the GIS spatial tools.

In accordance with the above, this paper evaluates the feasibility of a mechanistic modelling approach to field-based predictions of rainfed sugar-cane yields under various crop and environmental conditions.

MATERIAL AND METHODS

Sugar-cane growth was simulated in the 1995 cropping season in farms of Cuba, Mexico and Venezuela using the mechanistic model SWAP (Van Dam et al., 1997). SWAP is a physically based hydrological model, developed from the model SWACROP (Wesseling et al., 1989). The SWAP model is a functional combination of an advanced soil water simulation model and a crop-growth simulation model. The Richards equation for soil-water movement is solved in SWAP by a numerical scheme, including different practical options for the initial and boundary conditions. SWAP follows the same approach for crop-growth simulation as the model SUCROS (Spitters et al., 1989) and other more advanced Dutch models (Van Dam et al., 1997). According to this approach, the radiation energy absorbed by the crop canopy is a function of incoming radiation and crop leaf area. Using the absorbed radiation and taking into account photosynthetic leaf characteristics, the potential gross photosynthesis is calculated. The latter is reduced due to water stress, as quantified by the relative transpiration (i.e., the ratio between actual and potential transpirations), and yields actual gross photosynthesis. The amount of water that crop roots are able to extract from soil pores determines the relative transpiration ratio. SWAP uses the Feddes et al. (1978) root water uptake function for this calculation, which depends on the crop.

Ruiz et al. (1999) provided some sugar-cane functions, as needed in SWAP simulations. These functions comprise the fitted Feddes et al. (1978) root water-uptake function, the partitioning of soil evaporation and crop transpiration and the interception of rainfall by the sugar-cane canopy. Like Keating et al. (1999), several crop coefficients determined for maize (Van Dam et al., 1997) are considered valid for sugar-cane also. Furthermore, an interface between the GIS ILWIS (v 2.2) and SWAP, written in Delphis 4.0 is a visual-friendly code running in a Windows environment. The interface reads firstly the attributes from a GIS-exported table and writes the SWAP input files for each sugar-cane field. Secondly, the interface runs SWAP for each polygon and writes the final yield output (Van Dam et al., 1997) as in a GIS-readable file for each field unit.

The influence of water availability on sugar-cane growth is paramount for the first 150 days after sprouting, corresponding to the higher growth stage (Doorenbos and Kassan, 1986; Fernandez et al., 1990; Robertson et al., 1999; Wiedenfeld, 2000). So, SWAP simulations were carried out only for 150 days after sugar-cane emergence. Final yields (in t ha^{-1}) were calculated from the dry matter produced after the 150-day simulation period, using the calibration coefficient provided by Utset et al. (1999). Emergence dates were assumed to be 30 days after seeding or last harvest dates. The Penman-Monteith potential evapotranspiration was calculated daily and assumed as the topsoil boundary condition. Since no information was available regarding soil water movement at the bottom of the 3-m depth simulated soil layer, null water flux was taken as the bottom boundary condition. A water table at 250 cm depth in equilibrium with soil water was assumed as the initial condition. The spatial unit considered in the simulations is a sugar-cane field, i.e., a field showing the same cultivar, ratoon, emergence and harvest date as well as water and fertilizer management.

Compilation of basic information for the mechanistic-modelling evaluation in each country comprises the soil hydraulic properties in each sugar-cane field within the farms selected, as well as sugar-cane yields and emergence dates in each field and the daily mean values of wind velocity, maximum and minimum air temperatures, air humidity and sun net radiation. Since only one sugar-cane farm of several hundred hectares was considered for simulations, climate data were assumed to be spatially constant for the entire experimental zone in each country. Cultivars and ratoons in each field were also taken into account while comparing simulated and actual yields. Though the sugar-cane functions provided by Ruiz et al. (1999) do not consider the cultivar and ratoon effects in modelling results, actual sugar-cane yields depend on these features and hence the modelling performance accuracy could be affected by cultivar and ratoon variability. The cadastral maps of each sugar-cane farm (depicting the sugar-cane field distribution) were digitized and converted in ILWIS vector layers. Emergence dates and soil hydraulic properties were introduced in the GIS as spatial attributes of each polygon. The interface between the GIS and SWAP was run later to provide the simulated yield for each polygon.

Cuban Experimental Zone

A farm of the sugar-cane factory 'Héctor Molina' located at 22°44′ N, 82°10′ W, Havana province and situated 20 m above sea level, was selected as the experimental zone in Cuba. This farm contains 82 field units, although only 45 were cropped in the 1995 season. The total farm area is 1005 ha and three different soil types are present: Vertisols, Gleysols and Ferralsols. Seven sugar-cane cultivars and one ratoon were cropped in the season. Sugar-cane emergence dates range from December 1994 to April 1995, according to the seeding or last harvest dates. Sugar cane was cropped at the Cuban farm under rainfed conditions. The climate at the experimental site was characterized by a mean annual precipitation of 1349 mm and a mean annual evaporation (from class 'A' pan measurements) of 1977 mm. About 80% of the total precipitation falls from April to September. Mean annual temperature is 24°C, with maximum and minimum daily mean annual temperatures 29°C and 18.5°C respectively. The 1995 monthly precipitations are depicted in Figure 30.1A.

The soil hydraulic properties were calculated using the Rawls et al. (1982) as well as the Rawls et al. (1998) pedotransfer functions. The soil water content values of the soil-water characteristic curve and the saturated hydraulic conductivity of the sugar-cane root zone were determined as the average of two depths (0-30 and 30-60 cm) in seventeen sites in the Héctor Molina area, using the soil data (i.e., mechanical composition, soil density and organic matter) provided by the 1:25000 national soil survey (NDSF, 1985). The coefficients of the Van Genuchten (1980) model parameters (alpha, n, residual and saturated water contents) were estimated using code RETC (Van Genuchten et al., 1991). Raster maps for each of the Van Genuchten parameters were produced for the entire sugar-cane farm. Average values of these parameters were calculated and assigned to each sugar-cane field polygon through a GIS analysis, similarly to the procedure described by Richter (1999).

Mexican Experimental Zone

The ejidal part (small owners) of the sugar-cane factory Casasano, located at 18°50′ N, 98°57′ W in Cuautla, Morelos state, was the Mexican experimental zone. The ejidal comprises 96 small fields with a total area of 196 ha. The zone altitude varies from 1260 to 1320 m (north side) above sea level. Cambisols are the dominant soils in the zone, although Vertisols are also found. Six different sugar-cane cultivars and up to four ratoons were cropped in the Mexican experimental zone. Even though surface irrigation is occasionally provided to the ejidal, there is no information available about irrigation depth and application timing and hence this water supply was not considered. Sugar-cane fields seeded in January were excluded from this analysis since they had an irrigation supply. Accordingly, 73 fields cropping sugar-cane in 1995 were used for comparing actual and simulated yields. The climate in the zone is characterized by a total average annual precipitation of 912 mm. About 94% of the rainfall falls from May to October. Mean annual temperature is 20.5°C, with average maximum and minimum temperatures 31.7 and 15.2°C respectively. Figure 30.1B shows the monthly precipitation. Sugar-cane emergence dates extend from March to July.

Soil hydraulic properties were calculated through the same approach followed in the Cuban zone; they were estimated from the Rawls et al. (1982) and Rawls et al. (1998) pedotransfer functions. Data of soil mechanical composition, soil density and organic matter were available for three depths (0-25, 25-50 and 50-75 cm) at eighteen locations inside the experimental zone. As done for the Cuban farm, soil hydraulic properties calculation, saturated hydraulic conductivity and the Van Genuchten parameters were estimated at each of these locations

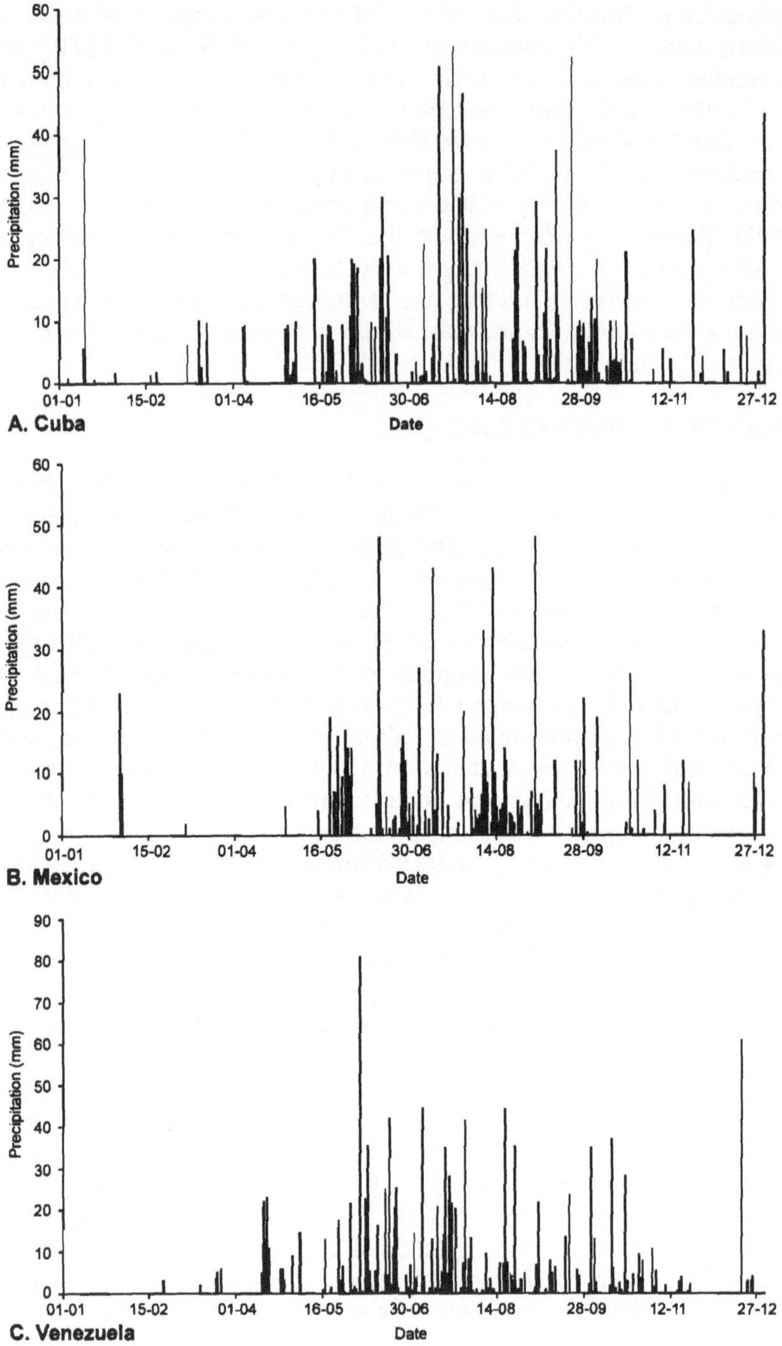

Fig. 30.1 Daily precipitation in 1995 in each of the experimental zones in Cuba, Mexico and Venezuela.

and raster maps created for each property by interpolation. Average values of the parameters were assigned to each of the 73 polygons by an ILWIS raster-vector analysis.

Venezuelan Experimental Zone

'El Paso', located at 9°30′ N and 69°40′ W in Turen, Portuguesa state, Venezuela, was the experimental zone in this country. The farm is located in the Venezuelan plains known as 'Los llanos' and covers a total area of 300 ha. Vertisol is the only soil type present in the farm. Six cultivars and up to three ratoons were cropped in 'El Paso' in the 1995 season. Like the Mexican farm, some irrigation is provided but not controlled at the field scale and was therefore ignored for present simulations. Likewise, sugar-cane yields corresponding to the fields seeded in December were not taken for comparison with the simulated yields because December-sprouted fields had been irrigated. Accordingly, sugar-cane emergence dates ranged from January to April and 62 fields were considered for this comparative study.

A mean temperature of 25°C and 1500 mm annual rainfall characterize the climate at the Venezuelan experimental zone, situated at an altitude of 200 m above the sea level. Precipitation behaviour in the considered period is shown in Fig. 30.1C. Soil hydraulic properties were not calculated through pedotransfer functions as these values were available for each sugar-cane field at the 0-30 cm depth. The Van Genuchten model parameters were estimated using RETC and directly assigned to the ILWIS polygon map.

Statistical Approach for Validating Model Results

Model validation analyses were done independently for each country. Mean error (ME) was calculated as a measure of simulation bias. Likewise, the root mean square error (RMSE) was related to simulation accuracy. RMSE values were obtained from (Smith et al., 1996):

$$RMSE = \frac{100}{\overline{AY}} \sqrt{\sum_{i=1}^{n} (SY - AY)^{e'} / n} \quad (1)$$

where AY are the actual yields, SY the simulated yields and n the number of yield measurements and simulations. Richter (1999) and Tubiello et al. (1999) used this approach for assessing their modelling performance accuracy.

Comparison of simulated and actual sugar-cane yields was done for all pairs of actual and simulated yields in each country. Mean *AY* and *SY* and their respective coefficients of variation, mean errors, as well as RMSE were calculated for all the data. These calculations were split according to emergence months, cultivars and ratoons. In addition, General Linear Models were fitted to *AY* and *SY* respectively for all three country data sets taken together. Sugar-cane cultivars, emergence months, ratoons and soil hydraulic properties were considered as the General Linear Model independent variables.

RESULTS

Cuba

Table 30.1 shows the mean, maximum and minimum values, standard deviations, coefficients of variation, medians, skewness and kurtosis of saturated water contents (SWC), saturated hydraulic conductivity (K_s), as well as the parameters α and n of the Van Genuchten model for the soil-water retention curve. According to Carsel and Parrish (1988), the average parameters of the soil hydraulic properties for the Cuban zone, (Table 30.1), correspond to a loam soil. However, notable differences between the maximum and minimum values of the soil hydraulic properties corresponding to the Cuban farm can be seen in Table 30.1. This means that the soil behaviour concerning water movement is quite different across the farm. The highest CV corresponds to the soil

Table 30.1 Mean values, maximum (MAX) and minimum (MIN) values, standard deviations (SD), coefficients of variation (CV), medians (MN), skewness (S) and kurtosis (K) of saturated soil-water content (SWC), saturated hydraulic conductivity (Ks), as well of alpha and *n* parameters of the Van Genuchten model for the soil-water retention curve in each country

		MEAN	*MAX*	*MIN*	*SD*	*CV*	*MN*	*S*	*K*
Cuba	SWC	0.60	0.64	0.56	0.03	4.4	0.60	-0.26	-0.92
	Ks	76.57	111.92	27.36	27.26	35.6	73.38	-0.24	-1.09
	alpha	0.04	0.06	0.03	0.01	21.0	0.04	-0.11	-0.89
	n	1.28	1.31	1.23	0.02	1.7	1.28	-0.58	-0.10
Mexico	SWC	0.62	0.70	0.55	0.04	5.9	0.62	0.44	-0.26
	Ks	175.47	292.00	52.00	75.63	43.1	158.00	0.01	-1.39
	alpha	0.05	0.09	0.02	0.02	29.1	0.05	0.53	0.07
	n	1.18	1.21	1.16	0.01	1.1	1.18	0.69	-0.34
Venezuela	SWC	0.49	0.56	0.43	0.03	6.6	0.49	0.13	-0.68
	Ks	24.30	31.20	9.60	8.82	36.3	31.20	-0.70	-1.22
	alpha	0.01	0.08	0.00	0.01	64.5	0.01	4.01	25.49
	n	1.26	1.55	1.23	0.03	2.4	1.25	8.41	82.16

hydraulic conductivity, among the other shown soil properties, as pointed out previously (Warrick and Nielsen, 1980). The small difference between the mean and the median, as well as between skewness and kurtosis values, suggests normal distributions for SWC and the parameter *n*. However, K_s and alpha spatial distributions do not seem to follow a normal probability law. According to Warrick and Nielsen (1980) and Vauclin (1982), K_s is not normally distributed very often, which accords with the present results.

Table 30.2 shows the Mean AY and SY and their coefficients of variation, mean errors and RMSE obtained in the Cuban experimental zone. The information is decomposed among emergence months (from January to April), the three most important cultivars and ratooning (sugar-cane plant and one ratoon). The average simulated yield is higher than actual yields. Accordingly, ME is negative in all the cases shown in Table 30.2 for the Cuban farm. The AY coefficients of variation (CV) are higher than SY CV. This means that actual yields are affected by other sources of variability, which are not taken into account by the model. Crop modelling is based on ideal conditions for crop growing. Hence, SY are very often higher than AY because the latter are affected by other factors.

Figure 30.2 shows the AY-SY comparison. Generally, SY are higher than AY for the Cuban zone (Fig. 30.2A). This is particularly evident for the lower values of actual yields. The correlation coefficient between all simulated and actual sugar-cane yields is 0.46, which is similar to the coefficients found in other studies (Jamieson et al., 1998; Rosenthal et al., 1998; Chipanshi et al., 1999; Tubiello et al., 1999). Likewise, Figure 30.3 depicts the mean AY and SY, as well as their standard deviations and the found RMSE. As can be seen in Figure 30.3, AY and SY are statistically equivalent in the Cuban farm, taking into account the corresponding standard deviations. Likewise, Figure 30.4 and Figure 30.5 depict AY and SY spatial distributions respectively. Sugar-cane yields below 50 t ha^{-1} were considered very low, from 50 to 80 low, and up to 100 t ha^{-1} acceptable; those of 100 to 150 were considered high and those higher than 150 t ha^{-1} very high yields. As can be seen by comparing Figures 30.4A and 30.5A, the model generally mimics AY spatial distribution in the Cuban farm. Nevertheless, though AY and SY correlate relatively well, large simulation errors were found performing the model in Cuba in accordance with the RMSE value shown in Table 30.2 and Figure 30.3. The RMSE value is even higher than the mean AY. It is known that average sugar-cane yields in Cuba are lower than the Cuban standard

Table 30.2 Mean actual (MAY) and simulated sugar-cane yields (MSY), standard deviation of actual (AYD) and simulated yields (SYD) and the respective coefficients of variation (AY-CV, SY-CV), mean error (ME) and root of the mean square error (RMSE) of the three Latin American countries. The information is decomposed according to the emergence date, cultivars and ratoons; accordingly the number of observations (N) and the ratio between overall and decomposed RMSE (RR) are also shown.

Cuba

	MAY	*AYD*	*AY-CV*	*MSY*	*SYD*	*SY-CV*	*ME*	*N*	*RMSE*	*RR*
All	60.50	49.49	81.8	91.69	23.84	26.0	– 31.2	45	87.6	1.00
Emergence										
Dec-Jan	44.19	26.02	58.9	47.91	0.77	1.6	– 3.7	4	52.0	0.59
Feb	42.74	19.37	45.3	82.52	7.51	9.1	– 39.8	18	102.0	1.16
Mar	48.48	19.83	40.9	100.69	1.63	1.6	– 52.2	6	113.4	1.29
Apr	87.39	69.60	79.6	108.53	24.42	22.5	– 21.1	17	75.6	0.86
Cultivar										
C-87-51	75.15	60.05	79.9	98.81	22.60	22.9	– 23.7	24	76.1	0.87
J-60-5	45.05	21.85	48.5	88.30	12.90	14.6	– 54.0	6	131.4	1.50
Ratooning										
Plant	42.32	21.01	49.6	91.29	14.54	15.9	– 36.1	4	93.0	1.06
Ratoon	63.85	52.61	82.4	97.61	20.25	20.7	– 33.8	38	88.8	1.01

(Table contd.)

(Table contd.)

Mexico

	MAY	AYD	AY-CV	MSY	SYD	SY-CV	ME	N	RMSE	RR
All	119.45	36.50	30.6	100.18	29.25	29.2	8.8	73	34.8	1.00
Emergence										
Mar	142.09	48.30	34.0	132.01	10.92	8.3	10.1	11	41.7	1.20
Apr	120.63	34.56	28.6	118.74	29.43	24.8	1.9	16	28.5	0.82
May	120.05	36.33	30.3	105.11	18.50	17.6	14.9	20	31.0	0.89
Jun	117.00	25.59	21.9	76.40	7.52	9.8	40.6	18	41.6	1.20
Jul	90.00	28.66	31.8	60.47	1.48	2.5	29.5	8	43.8	1.26
Cultivar										
CP 72-86	119.83	26.84	22.4	92.33	25.99	28.1	6.3	12	40.3	1.16
MEX 57-473	121.54	43.39	35.7	93.17	30.80	33.1	10.2	24	33.0	0.95
MY 55-14	115.94	34.52	29.8	78.57	27.83	35.4	28.1	39	34.6	0.99
Ratooning										
Plant	151.33	39.41	26.0	117.33	25.32	21.6	9.2	9	27.2	0.78
Ratoon	118.00	36.16	30.6	103.94	26.03	25.0	10.9	11	42.3	1.22
$Ratoon^2$	114.34	33.93	29.7	96.48	29.80	30.9	22.7	53	35.4	1.02

(Table contd.)

(Table contd.)

Venezuela

	MAY	*AYD*	*AY-CV*	*MSY*	*SYD*	*SY-CV*	*ME*	*N*	*RMSE*	*RR*
All	85.69	21.84	25.5	98.21	19.54	19.9	–12.5	62	27.3	1.00
Emergence										
Jan	64.86	21.97	33.9	64.58	12.37	19.2	0.3	10	20.4	0.75
Feb	100.49	16.01	15.9	109.84	9.14	8.3	–9.4	26	20.4	0.75
Mar	78.52	19.14	24.4	104.44	13.33	12.8	–25.9	19	41.9	1.53
Apr	79.93	9.26	11.6	86.11	2.07	2.4	–6.2	7	12.9	0.47
Cultivar										
C-323-68	102.13	14.83	14.5	99.87	15.44	15.5	3.7	13	16.4	0.60
C-371-67	97.31	19.06	19.6	112.73	6.37	5.7	–13.3	21	19.8	0.73
MY-5514	71.92	11.09	15.4	106.22	11.12	10.5	–28.1	9	47.3	1.73
V-756	78.94	9.47	12.0	86.36	2.13	2.5	–7.4	11	15.2	0.56
Ratooning										
Plant	100.94	16.39	16.2	107.99	12.48	11.6	–5.8	32	17.9	0.65
Ratoon 3	77.64	9.28	12.0	89.22	8.08	9.1	–23.7	13	40.2	1.47
Ratoon 4-6	63.12	13.21	20.9	86.66	26.56	30.7	–16.7	17	40.1	1.47

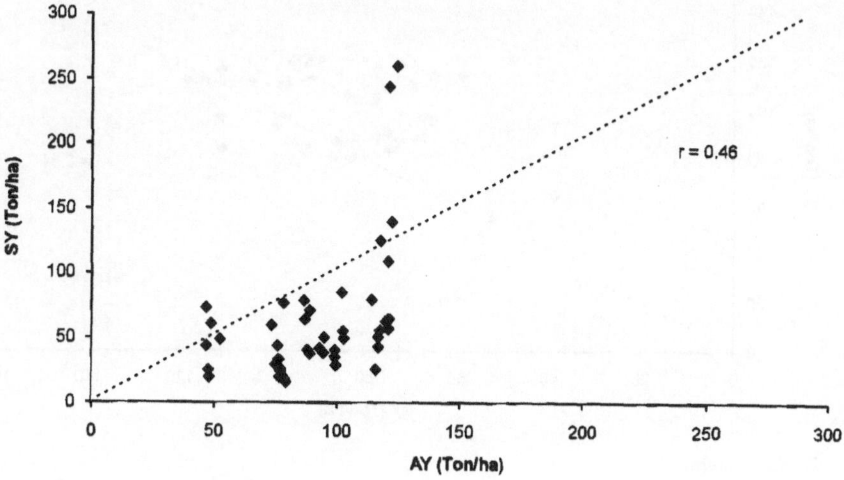

A. Cuba

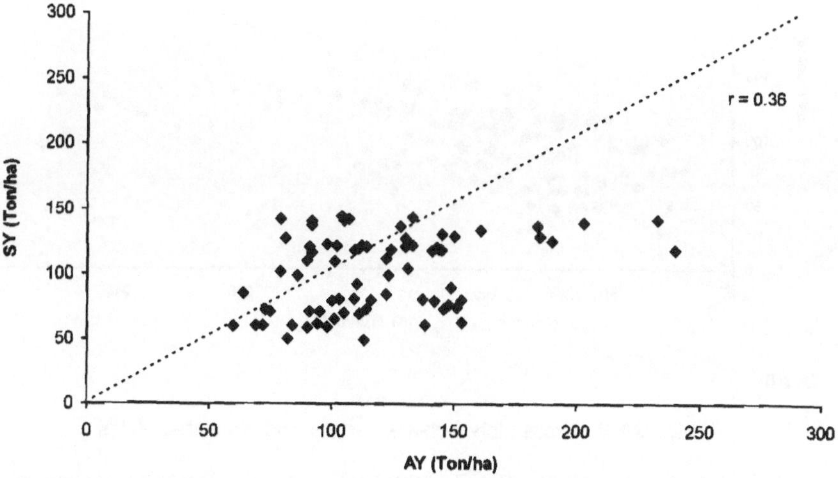

B. Mexico

Fig. 30.2

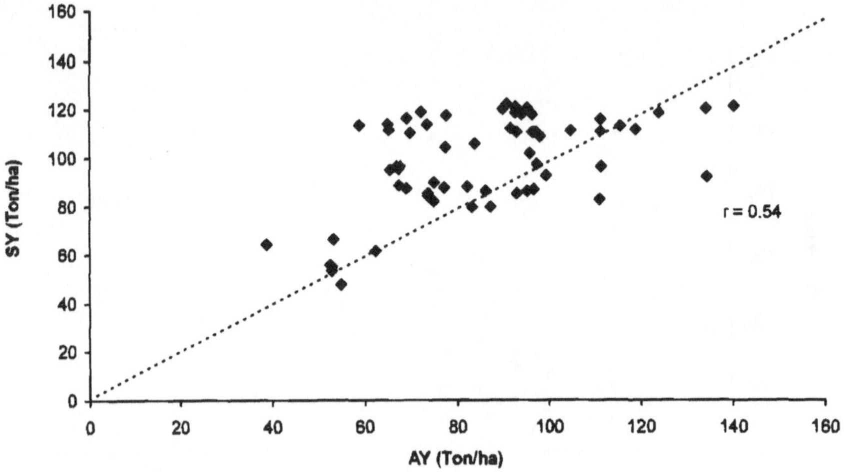

C. Venezuela.

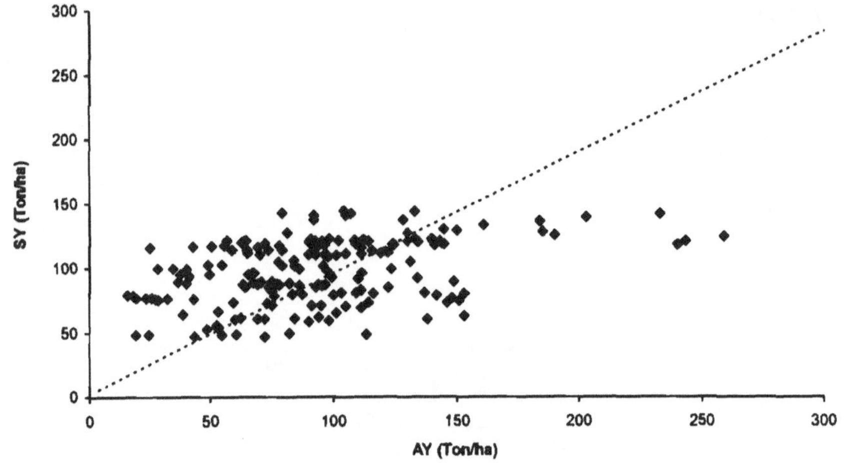

D. All

Fig. 30.2 Cross-plots between actual and simulated yields

yields during the 1990s (FAO, 1996), due to the serious shortage of fertilizers and other inputs after the political changes in Europe at the end of the 1980s. As outlined above, SWAP modelling assumes an 'ideal' growing, unaffected by other factors rather than water use. Hence, the modelling inaccuracies are high in this case. Likewise, Figure 30.6 depicts AY and SY means for each emergence month, as well as their

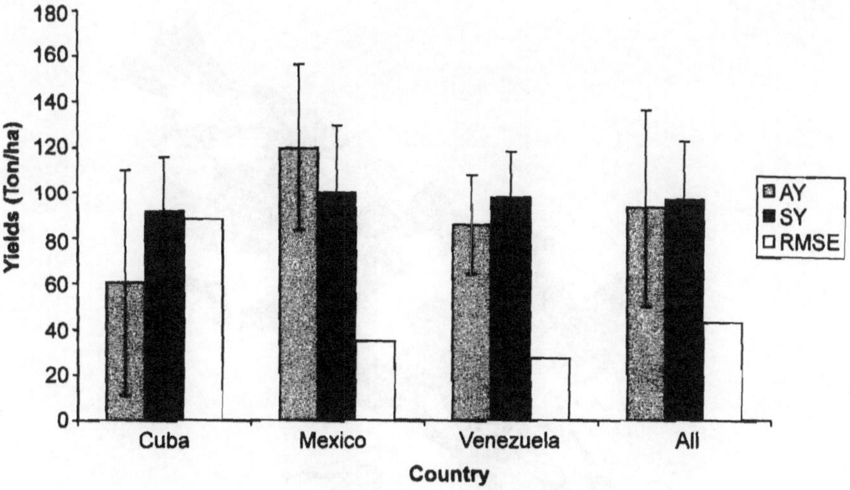

Fig. 30.3 Mean actual and simulated yields and corresponding RMSE. Y-bars show the standard deviations.

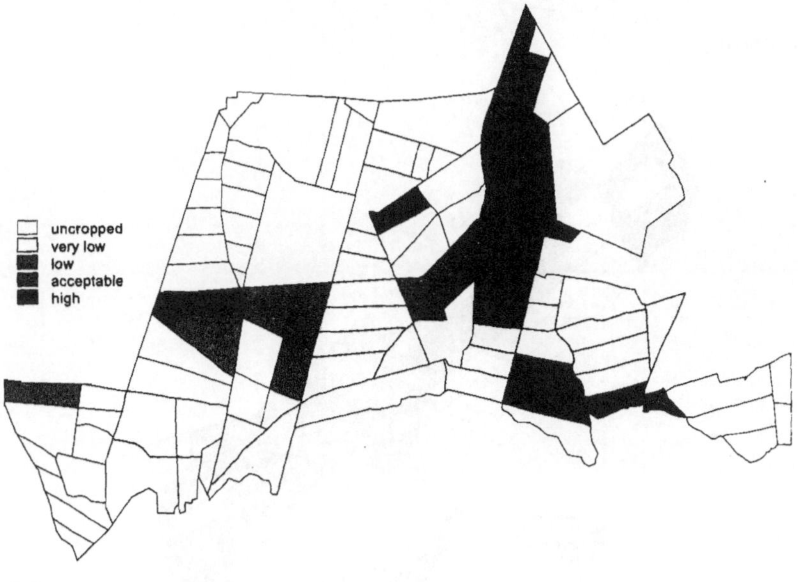

A. Cuba

Fig. 30.4

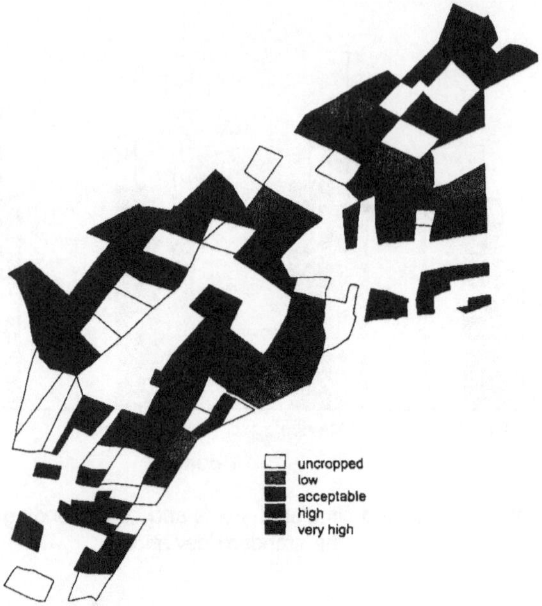

B. Mexico

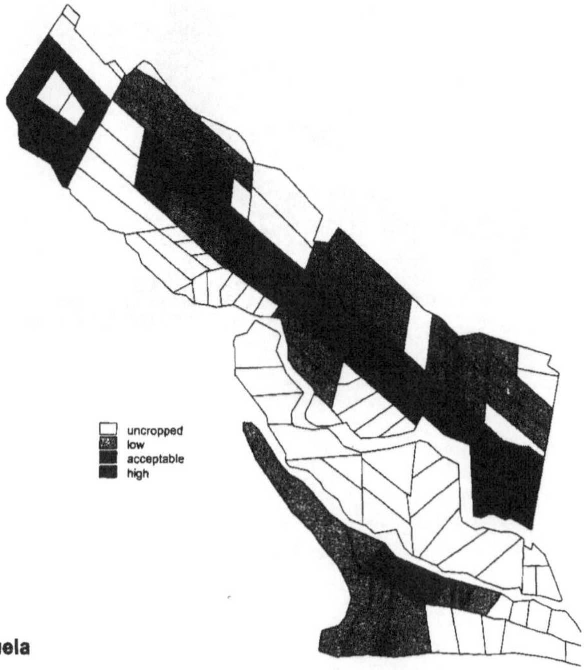

C. Venezuela

Fig. 30.4 Actual yields in each country

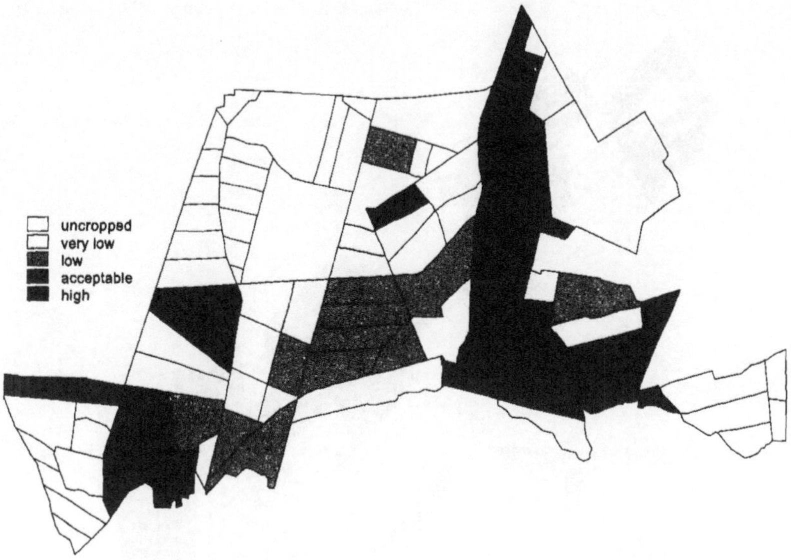

A. Cuba

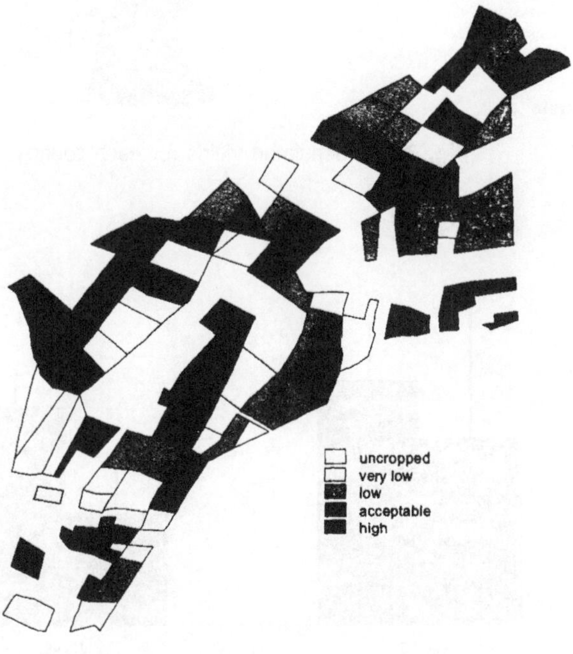

B. Mexico

Fig. 30.5

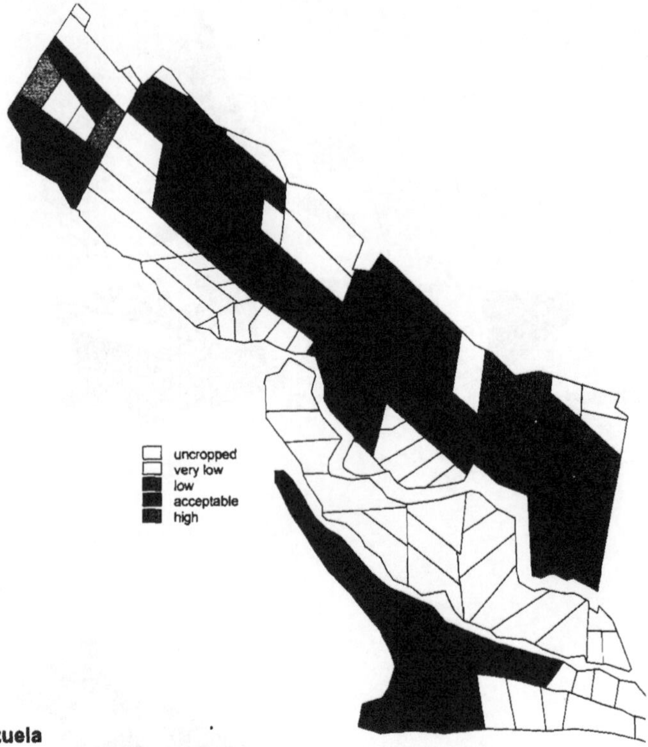

C. Venezuela

Fig. 30.5 Simulated yields for each country.

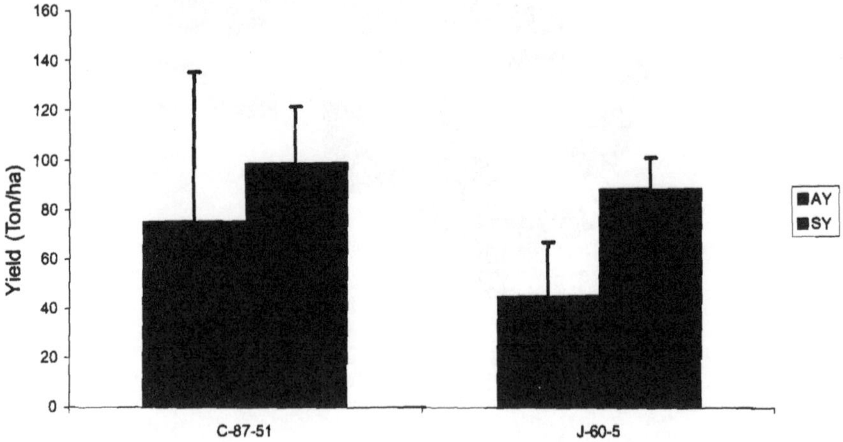

A. Cuba

Fig. 30.6

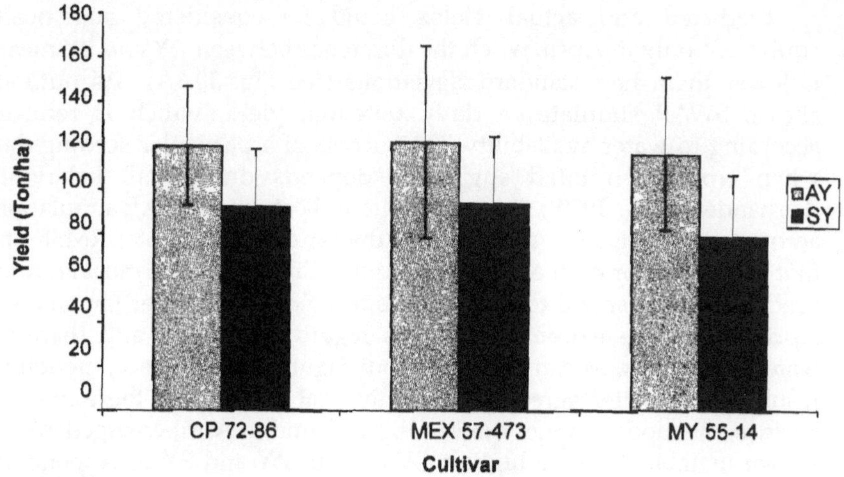

B. Mexico

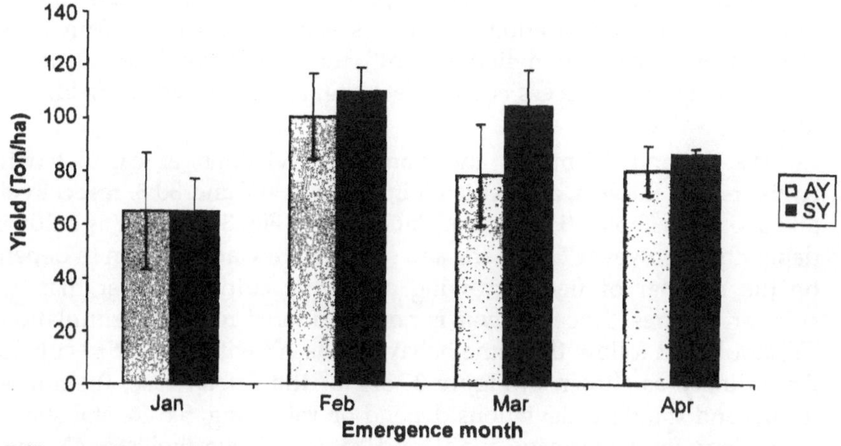

C. Venezuela

Fig. 30.6 Actual and simulated yields averaged by sugarcane emergence month. Y-bars show the standard deviations.

standard deviations. As can be seen in Figure 30.6A, both AY and SY are related to the emergence date in the Cuban farm. In addition, SY behaviour mimics AY distribution according to sugar-cane emergence month. The highest AY and SY correspond to sugar-cane sprouted in April. As can be seen in Figure 30.1A, rainfall becomes sufficient for sugar-cane growing after April, which corresponds to the usual conditions for rainfed sugar-cane cropping in Cuba (Fernandez et al., 1990).

Predicted and actual yields could be considered statistically equivalent only in April, when the difference between AY and SY means is lower than their standard deviations (see Fig. 30.6A). As outlined above, SWAP simulates a daily potential yield, which is reduced according to water availability. The success of a particular seeding date when cropping rainfed sugarcane depends on rainfall behaviour (Fernandez et al., 1990); present results indicate that SWAP simulations account for this effect. According to the ratio between total RMSE and that calculated for each emergence month (Table 30.2), the most accurate predictions correspond to sugar-cane sprouted in December-January and April. Those are extreme situations regarding rainfall and therefore water availability, as can be noted from Figure 30.1A. Hence, modelling results are in better agreement with the real situation in these months. Besides, the model mimics the variability found in April-emerged AY. As shown in Table 30.2, the highest CV of both AY and SY corresponds to this emergence month. The higher CV in this particular month could be related to the effect of soil hydraulic properties spatial variability. According to the SWAP simulation approach, higher soil water contents influence crop transpiration as well as water stress (Van Dam et al., 1997). Hence, oxygen deficit in both the actual conditions and the simulated growth process could affect sugar-cane growth in fields with lower K_s values.

The AY and SY means and standard deviations, as calculated for cultivars and ratoons, are depicted in Figures 30.7 and 30.8, respectively, and shown in Table 30.2. For the Cuban farm (Fig. 30.7A), AY is cultivar-dependent. However, SY means and standard deviations seem to depend on the number of fields growing a specific cultivar rather than the cultivar features. Since cultivar is not considered in SWAP simulations, SY should not follow the same behaviour as AY with respect to cultivar. Accordingly, as shown in Figure 30.8A for the Cuban case, though AY means and standard deviations depend on ratooning, SY are statistically equivalent for sugar-cane plant and ratoon. Nevertheless, AY, even depending on cultivar and ratooning, show very high standard deviations. This is particularly evident for C-87-51 and ratoon, which are the most frequent cases in the Cuban experimental zone (see Table 30.2). This means that cultivar and ratooning influence on AY is not very high compared to other factor effects.

Mexico

As shown in Table 30.1, soil hydraulic properties are higher in the Mexican farm than those obtained in Cuba. According to Carsel and

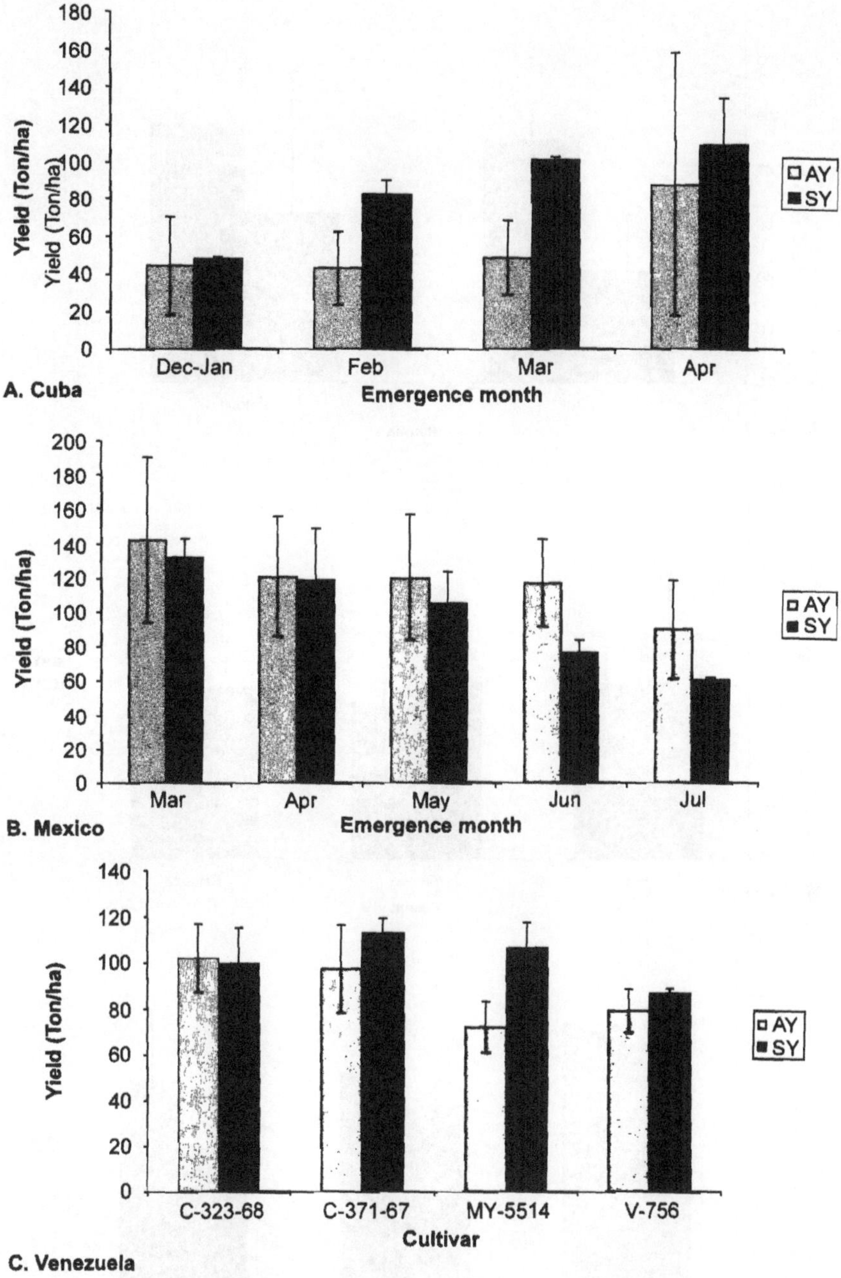

Fig. 30.7 Actual and simulated yields averaged by cultivar. Y-bars show the standard deviations.

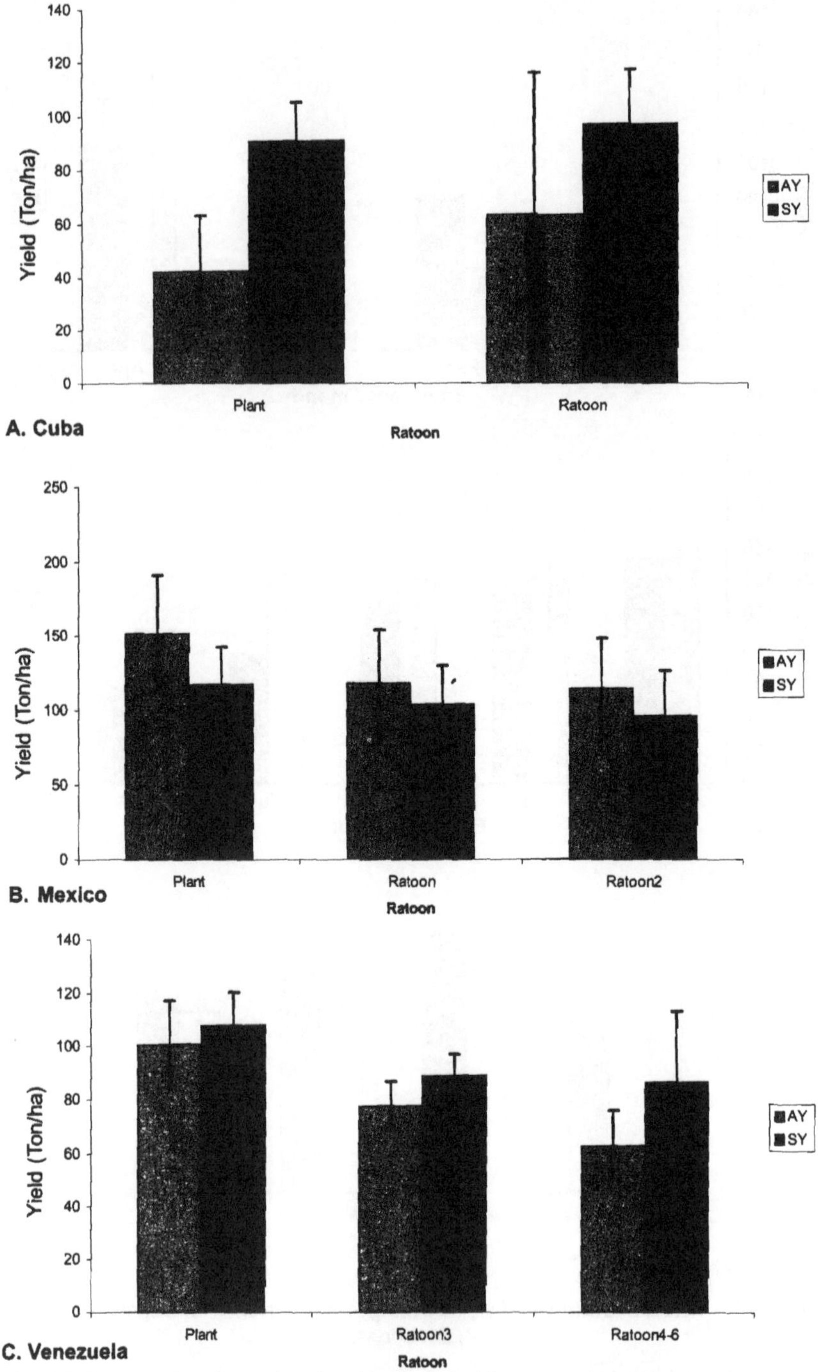

Fig. 30.8 Actual and simulated yields averaged by ratoon Y-bars show the standard deviations.

Parrish (1988), the average values of the soil hydraulic properties for the Mexican farm (Table 30.1), correspond to a sandy loam soil. The distribution probability of soil hydraulic properties in the Mexican experimental zone is similar to that previously shown in the Cuban farm. However, the hydraulic properties CV are higher in the Mexican farm than in the Cuban. Hence, the spatial variability of the soil hydraulic properties, one of the most important sources of error in mechanistic model performances, is relatively higher in the Mexican experimental zone.

The Mexican mean AY, (Table 30.2), are very high yields, compared to the internationally obtained and the average yields achieved in Mexico (FAO, 1996). Accordingly, as can be seen in Table 30.2, AY are higher than SY and ME is positive. The same is evident in the AY-SY comparison depicted in Figure 30.2B and Figure 30.3. Hence, the model underestimates the sugar-cane yields in the Mexican zone. The difference between AY and SY CVs is not so high as that found in the Cuban farm. On the other hand, the actual yield CV obtained in the Mexican experimental zone is lower than the corresponding CV found in the Cuban farm. As already mentioned, sugar-cane yields are affected by several uncontrolled factors, which could increase yield variation from one field to another. Conversely, the sugar-cane yields obtained in the Mexican experimental zone are extremely high. Hence, it may be supposed that most of the constraints affecting sugarcane growth in the Cuban farm were precluded in the Mexican field units. Furthermore, irrigation was supplied to the ejidals of the Mexican farm. Since this supply was not controlled, it was not taken in account in this analysis. Therefore, the model underestimation of sugar-cane yields could also be due to this factor.

As can be seen in Figure 30.2B, the model underestimates the sugar-cane yields in many fields. The total correlation coefficient is 0.36, which is relatively low compared to the coefficient obtained in the Cuban farm, but higher than those reported by Landau et al. (1998). In particular, the model inefficiently predicts yields higher than 150 t ha^{-1}. Actually, the sugar-cane yields obtained in the Mexican zone are quite unusual. Extremely smaller sugarcane fields (average area 2 ha), belonging to different owners, allow the farmers to take better care of their fields than the Cuban farmers can. Nevertheless, despite the model predictions inability to estimate the Mexican higher yields, it still does mimic the AY spatial distribution, as can be seen on comparing Figures 30.4B and 30.5B. Besides, mean RMSE is lower than that obtained for the Cuban farm, indicating that, on average, SY predictions are accurate enough.

Figure 30.6B depicts the average actual and simulated yields for each month of emergence, as well as their standard deviations. Unlike in the Cuban case, AY decreases from March to June. However, like the Cuban results, the model explains the changes in yields according to the seeding month. Moreover, according to the obtained AY standard deviations, AY and SY could be considered statistically equivalent for most of the studied months. The dependence of AY and SY on cultivars is shown in Figure 30.7B for the Mexican farm. Similar to the Cuban case, the cultivar influence in AY is not evident. The AY and SY means and standard deviations separated by cultivar are quite similar. Nevertheless, unlike the Cuban results, AY for sugar-cane plant are higher than those obtained for ratoons. They agree with the normal reported case (Fernandez et al., 1990). The SY behaviour with respect to ratoon is similar to the actual behaviour, even though the model does not simulate ratooning effects. The Ruiz et al. (1999) sugar-cane functions included in SWAP simulations were developed for the sugar-cane plant. Hence, the model could be mimicking sugarcane plant growth rather than ratoons. Accordingly, the model performing relative accuracy, as characterized in Table 30.2 by the ratio between ratooning-calculated RMSE and total RMSE for the Mexican case, is considerably lower for the sugar-cane plant than for ratoons.

Venezuela

According to the results shown in Table 30.1, the Venezuelan K_s average is relatively low. This value, as well as most of the parameters of the Van Genuchten model shown in Table 30.1 for Venezuela, correspond to a clay soil (Carsel and Parrish, 1988). The spatial variability of SWC and n seems to be normally distributed, similar to the results found in Cuba and Mexico. Accordingly, the highest CVs correspond to K_s and the parameter alpha. The Venezuelan soil hydraulic properties generally show the highest CV among all those shown in Table 30.1; only K_s CV in the Mexican farm is higher than the Venezuelan CV. Nevertheless, no reliable comparison can be made in this case since soil hydraulic properties in Venezuela were obtained from direct measurements in each sugar-cane field, whereas in Cuba and Mexico they were estimated from pedotransfer functions and smoothened through GIS interpolation and polygon-averages.

As can be seen in Table 30.2, the average actual sugar-cane yields are higher in the Venezuelan farm than those found in Cuba, but lower than those obtained in Mexico. The Venezuelan farm, as well as the surrounding land, belongs to a single owner and workers are contracted for farm labour. This form of land property and sugar cane production is the most common in Latin America and other places in the world. In fact, as mentioned earlier the average yields achieved in the Venezuelan zone accord with the average sugar-cane yields reported internationally and those obtained in Cuba during the 1980s (FAO, 1996), corresponding to sugar-cane cropping under sufficient fertilizers and other supplies. AY and SY in the Venezuelan farm are compared in Figure 30.2C. Like the simulations done for the Cuban farm, SY are generally higher than AY. This is due to the inability of the model to take into account factors that can affect yields in the real life, as explained above. Accordingly, the ME shown in Table 30.2 are negative. However, similar to the results obtained for the Mexican farm, the model inaccurately predicts higher yields. This means that it overestimates the effect of water stress (or water excess) in these cases. The correlation coefficient is 0.54, the highest obtained in the present study results. This correlation coefficient could be considered very good, given that comparison between AY and SY was done for productive conditions and that many uncontrolled factors affected AY. Rosenthal et al. (1998) found similar correlation coefficients when comparing actual and simulated sorghum yields in large regions of Australia. Their found correlations were considerably lower than those obtained in experimental plots, under controlled conditions. Furthermore, RMSE for the Venezuelan farm are lower than those obtained in Cuba and Mexico, which means that the best model performance was achieved in the Venezuelan case.

As obtained in the Mexican farm, the coefficients of variation of both AY and SY are very close, although AY shows higher variability. This indicates that even though the model cannot reproduce all the yield-affecting factors, it can mimic yield variability. In accordance with this, the AY and SY spatial distributions depicted in Figures 30.4A and 30.5A, show relatively good agreement.

Figure 30.6C shows AY and SY means and standard deviations as calculated for emergence month. As can be seen from this Figure and noted from Table 30.2, the model mimics the emergence date effects on AY. The differences between AY and SY means are within the range of AY standard deviations and hence AY and SY could be considered statistically equivalent for all the emergence months. Furthermore, contrary to the Cuban and Mexican results in which AY increases or decreases with the emergence month, in the Venezuelan case the average AY of sugarcane sprouted in January are lower than in February, decreasing in March and April. As can be seen in Figure 30.6C, SY mimics this behaviour exactly. The AY and SY means and standard deviations calculated for each cultivar are depicted in Figure 30.7C. As expected, SY are generally unable to follow the AY dependence on cultivar. However, the differences between AY and SY for each particular cultivar are frequently lower than AY standard deviations and AY differences among cultivars. A very good agreement between AY and SY was found for C-323-68. The Ruiz et al. (1999) functions were experimentally obtained for this cultivar in Cuba. This could explain this good accordance. Correspondingly, the AY-SY agreement, as shown in Figure 30.8C and Table 30.2, is higher for sugarcane plant than for ratoons. However, the different factors affecting sugarcane yields, decomposed in Table 30.2, could interact among each other, which makes it difficult to assess the effects of a single factor. Besides, as was pointed out above, the model does not take into account ratoon or cultivar. Hence, any attempt to explain SY behaviour according to these crop features is purely speculative.

Overall Data Analysis

A comparison between all AY-SY pairs is shown in Figure 30.2D. As can be seen from this Figure, a linear relationship still exists between AY and SY, even considering the three-country data. The correlation coefficient is 0.41, which is higher than the coefficient obtained for the Mexican farm. Correspondingly, the difference between AY and SY means, as depicted in Figure 30.3, is lower than those obtained for each country. However, the SY standard deviations, as well as RMSE, are higher than those found in Mexico and Venezuela. This means that, as expected, the overall data analysis is affected by the bad model performance in the Cuban farm.

Table 30.3 shows the variance analysis of GLM results for both actual and simulated yield regressions. As can be seen from the Table, sprouting month is highly significant for actual and simulated yields (very low p values). The dependence of actual sugar-cane yields on emergence-sprouting month in the rainfed case is due to precipitation distribution and other climate-related factors. Accordingly, simulated yields are strongly affected by sprouting month. This means that precipitation distribution influences mechanistic modelling estimations as well as in real life. Besides, soil hydraulic properties are significant variables in both GLM regressions. In particular, according to the results shown in Table 30.3, SWC and K_s exhibit the higher influence in actual and simulated yields. Parameter alpha is related to the slope of the soil-water retention curve and it shows a very strong influence on simulated yields. As pointed out above, soil hydraulic properties were calculated by different methods in each country. Hence, no reliable comparison analysis of the specific influence of each property on simulated or actual yields could be done. The cultivar is non-significant in both GLM. On the contrary, ratooning seems to affect both actual and in particular simulated yields. However, the ratooning influence could be due to some cross-interaction among variables.

CONCLUSIONS

The mechanistic model for estimating sugar-cane yields under variable climate and soil conditions predicts the actual yields with reasonable accuracy. The model predictive ability is lower in zones where actual yields are extremely low or extremely high. Nevertheless, the model performs relatively well in the zones where yields are about 80-90 t ha^{-1} and unfavourable climate conditions are the main constraints for achieving higher yields. Moreover, the model accurately mimics the climate-based yield variation, as well as the effects of emergence or commencement of growth period in a particular month.

The accuracy of the model outputs depends on the cultivar and the ratoons. However, since the present version of the model does not take these features into account, the variability of the model results depends on neither cultivar nor on ratoons, except for the specific cultivar and ratoon considered in the experimentally developed sugar-cane functions used in the mechanistic modelling performances. Soil hydraulic properties, on the other hand, considerably influences model results. These properties range from badly drained clayey soils to sandy soils in the experimental zones. Even though the soil hydraulic properties within each experimental zone show a considerably spatial variability, the model performed relatively accurately in mimicking the spatial distribution of sugar-cane yields. Hence, the GIS approach for field-based interpolations of soil hydraulic properties can provide relatively accurate estimations of these properties.

The model result accuracy is site-specific. Many factors besides climate and soil types can influence sugar-cane yields. These include crop management, presence of diseases or weed infestations, as well as relatively remote motives such as political relationships among states or land property. Therefore, modelling should be used locally for assessing crop trends under several water managements instead of for predicting yields.

Acknowledgements

The studies presented in this paper were supported by the Inter-American Institute for Global Change Research (IAI), Initial Scientific Program, Round III, Proposal 41.

References

Burke, S., Mulligan, M. and Thornes, J.B. 1999. Optimal irrigation efficiency for maximum plant productivity and minimum water loss. Agric. Water Management 40: 377-391.

Carsel, R.F. and Parrish, R.S. 1988. Developing joint probability distributions of soil water retention characteristics. Water Resour. Res. 24: 755-769.

Connolly, R.D. 1998. Modelling effects of soil structure on the water balance of soil-crop systems: A review. Soil Till. Res. 48: 1-19.

Chipanshi, A.C., Ripley, E.A. and Lawford, R.G. 1999. Large-scale simulation of wheat yields in a semi-arid environment using a crop-growth model. Agric. Systems 59: 57-66

Doorenbos, J. and Kassan, A.H. 1986. Yield response to water. FAO. Irrigation and drainage. Paper No.33, 193 pp.

FAO (Food and Agriculture Organization). 1996. Year Book Production (49), FAO, Rome, 236 pp.

Feddes, R.A., Kowalik, P. and Zaradny, H. 1978. Simulation of Field Water Use and Crop Yield. Simulation Monographs, Pudoc, Wageningen, The Netherlands, 189 pp.

Fernández, A.R.; Dávila, A.I. and del Toro, M.F., 1990. Botánica y fisiología de la caña de azúcar. Pueblo y Educación, La Habana, 247 pp.

Fonseca, R. 1984. Necesidades de agua de la caña de azúcar plantada en diferentes épocas de siembra en el Occidente de Cuba. Ph. D. Diss., Havana.

Gabrielle, B., Menasseri, S. and Houot, S. 1995. Analysis and field evaluation of the Ceres models water balance component. Soil Sci. Soc. Amer. J. 59: 1403: 1412.

Jamieson, P.D., Porter, J.R., Goudriaan, J., Tirchie, J.T., van Keulen, H. and Stol, W. 1998. A comparison of the models AFRCWHEAT2, CERES-Wheat, Sirius, SUCROS2 and SWHEAT from wheat grown under drought. Field Crops Res. 55: 23-44.

Keating, B.A., Robertson, M.J., Muchow, R.C. and Huth, N.I. 1999. Modelling sugarcane production systems. I. Development and performance of the sugarcane module. Field Crop Res. 61: 253-271.

Landau, S., Mitchell, R.A.C., Barnett, V., Colls, J.J., Craigon, J., Moore, K.L. and Payne, R.W., 1998. Testing winter wheat simulation models' predictions against observed UK grain yields. Agric. For. Meteorol. 89: 85-99.

Leenhardt, D., Voltz, M. and Rambal, S. 1995. A survey of several agroclimatic soil water balance models with reference to their spatial application. Eur. J. Agron. 4 (1): 1-14.

Maraux, F. and Lafolie, F. 1998. Modeling soil water balance of a maize-sorghum sequence. Soil Sci. Soc. Amer J. 62: 75-82.

NDSF (National Direction of Soil and Fertilizers). 1985. Mapa de suelos de Cuba a escala 1: 25000. Ministerio de la Agricultura, La Habana.

Rawls, W., Brakensiek, D. and Saxton, K. 1982. Estimation of soil water properties. Trans. ASAE. 25: 51-66.

Rawls, D., Gimenez, D. and Grossman, R. 1998. Use of soil texture, bulk density and slope of the water retention curve to predict saturated hydraulic conductivity. Trans ASAE, 41: 983-988.

Reynoso, A. 1862. Ensayo sobre el cultivo de la caña de azúcar. La Habana, 462 pp.

Richter, G.M. 1999. Variability of winter rye grain yield in a glacial plain catchment – modeling and observation. Europ. J. Agron. 11: 239-253.

Rivero, L. 1985. Régimen hídrico de los suelos Ferralítico Rojo típicos de la región de Jovellanos, Cuba. Ph. D. Diss., Havana.

Robertson, M.J., Inman-Bamber, N.G., Muchow, R.C. and Wood, A.W. 1999. Physiology and productivity of sugarcane with early and mid-season water deficit. Field Crops Res. 64: 211-227.

Rosenthal, W.D., Hammer, G.L. and Butler, D. 1998. Predicting regional grain sorghum production in Australia using spatial data and crop simulation modelling. Agric. For. Meteorol. 91: 263-274.

Ruiz, M., Utset A., Medina, H., García, J., Fleitas, E., Ruiz, J., and Ferrer, M. 1999. Modelo determinístico para estimar el rendimiento de la caña de azúcar. Research Report 050130016-11, National Program for Global Change Research, Ministry of Sciences, Havana.

Singh, R., Subramanian, K. and Refsgaard, J.C. 1999. Hydrological modeling of a small watershed using MIKE SHE for irrigation planning. Agric. Water Management 41: 149-166.

Smith, J., Smith, P. and Addiscot, T. 1996. Quantitative methods to evaluate and compare soil organic matter models. In: D.S. Powlson et al. (eds.) Evaluation of Organic Soil Models: Using Existing Longterm Datasets. NATO ASI Series I, vol. 38. Springer-Verlag, Berlin.

Spitters, C.J.T., van Keulen, H. and van Kraalingen, D.W.G. 1989. A simple and universal crop growth simulator: SUCROS87, pp. 147-181. In: R. Rabbinge, S.A. Ward and H.H. van Laars (eds.). Simulation and Systems Management in Crop Protection. Simulation Monographs, Pudoc, Wageningen, The Netherlands.

Thompson, G.D. 1977. Irrigation of sugarcane. S. African Sugar J. 61: 126-131.

Tubiello, F.N., Rosenzweig, C., Kimball, B.A., Pinter, P.J., Wall, G.W., Hunsaker, D.J., LaMorte, R.L. and García, R.L. 1999. Testing CERES-Wheat with free-air carbon dioxide enrichment (FACE) experiment data: CO_2 and water interactions. Agron. J. 91: 247-255.

Utset, A., López, J. and Alvarez, M. 1999. Inter-American Institute for Global Change Research (IAI) Proposal ISP-III 41 "Estimation of ENSO effects on sugarcane yields in several Latin American countries". Annual Report.

Van Dam, J.C., Huygen, J., Wesseling, J.G., Feddes, R.A., Kabat, P., van Walsum, P.E.V., Groenendijk, P. and van Diepen, C.A. 1997. Theory of SWAP version 2.0. Report 71. Technical Document 45, Wageningen, The Netherlands, 167 pp.

Van Genuchten, M. 1980. A closed-form equation for predicting the hydraulic conductivity of unsaturated soil. Soil Sci. Soc. Amer J. 44: 892-898.

Van Genuchten, M., Leij, F. and Yates, S. 1991. The RETC Code for quantifying the hydraulic functions of unsaturated soils. US Sal. Lab., CA.

Vauclin, M. 1982. Méthodes d'etude de la variabilité spatiale des propietés d'un sol. In: Proceedings of Variabilité spatiale des processus de transfert dans les sols, Avignon.

Warrick, A. and Nielsen, D.R., 1980. Spatial variability of soil physical properties in the field, pp. 319-14. In: D. Hillel (ed.). Applications of Soil Physics. Academic Press, N.Y.

Wesseling, J., Kabat, P., Van Den Broek, B. and Feddes, R. 1989. Simulation function of water balance of a cropped soil with different types of boundary conditions including the possibility of drainage and irrigation and the calculation of crop yield (SWACROP). Instructions for input. WSC for Integrated Land, Soil and Water Research, Wageningen, the Netherlands.

Wiedenfeld, P.R. 2000. Water stress during different sugarcane growth periods on yield and response to N fertilization. Agric. Water Management 43: 173-182.

31

RESOURCE USE EFFICIENCY OF *COCHORUS, CELOSIA* AND *AMARANTHUS* AMONG FARMERS IN OYO STATE, NIGERIA

K.O. Adenegan[1] and O.I. Oladele[2]
[1]University of Ibadan, Department of Agricultural Economics, Nigeria
[2]University of Ibadan, Department of Agricultural Extension and Rural Development, Nigeria
e-mail: oladele20002001@yahoo.com

INTRODUCTION

Vegetables are plants consumed in relatively small quantities as a side dish or a relish with staple food. Most vegetables are the leaves, roots or stems of herbaceous plants, although flowers, calyces, immature seeds or fruits may also be eaten as vegetables. Vegetables constitute a rich and comparatively cheap source of fibre, protein, minerals, vitamins and calories for a large sector of mankind.

On the other hand, Rice and Colleagues (1993) have described vegetables as plants that provide a source of food often low in calories and dry matter content, eaten along with a basic starch food to make it more palatable. Three major classes of vegetables are consumed in Nigeria: a) those gathered from the wild such as leaves of purslane (*Portulaca*); b) indigenous vegetables, often gathered but also cultivated such as *Amaranthus* and *Celosia;* and c) imported vegetable species that are also cultivated.

According to FAO (1998), the basic daily per capita requirement for vegetables is 157 g, consisting of 47 g obtained from leaves and 110 g from fruit vegetables, pulses, roots and tubers. However, there is a wide gap between the required (157 g) quantity of vegetables and their availability (87.9 g) according to the Sri Lanka Department of Census and Statistics (1982). Leafy vegetables in particular are nutritionally important. They contain protein, minerals and vitamins (A in particular) and provide roughage. According to an annual report of the International Institute of Tropical Agriculture, a study of 150 urban households showed that for the poorer people, leafy green vegetables generally supply the protein not affordable in meat; they are the greatest source of calcium, iron, thiamine, riboflavin and vitamin C for both the poor and better off.

Vegetables are also a source of roughage, providing an indigestible matrix, stimulating muscles and preventing constipation through their laxative effect. Some vegetables have a curative action. All play a vital role in the health and nutrition of people throughout the world while providing employment and income that contribute to a higher standard of living. Nigeria is no exception.

Because they are short-duration crops, vegetables fit well into any cropping system and it is not uncommon in Nigeria to harvest 3-4 crops in the same field annually. Vegetables give higher tonnage per unit area in less time than cereal and offer better food value. Due to its labour intensiveness, vegetable production also offers better employment opportunities and increases farmer income. A certain quantity of bulky food is necessary for good health and leafy vegetables serve this function well when taken in sufficient quantity.

Sources other than the FAO have recommended per capita consumption per day of leafy vegetables ranging from 85 to 114 g and of non-leafy from 57 to 113 g. In other words, between 142 and 227 g of vegetables are thought to be needed every day. Further, an additional 57 g of vegetables daily has been recommended when fruit cannot be included in the diet.

The site for vegetable production is determined mainly by physical (climate and soil), economic and sociological factors. The physical factors play a major role in determining which species of vegetables can be grown and the potential productivity of a particular site.

Socio-economic factors can modify decisions regarding site choice, in densely populated areas for instance, or limit selection in other ways of a suitable site for growing vegetables.

Nonetheless, vegetables remain the primary source of protein, minerals and vitamins in developing countries, especially for the lower income groups. It is therefore important to seriously encourage their production and consumption, in particular those familiar to the community. Concomitantly the stages of development and growth and the various forms of vegetable preparation for optimum consumption benefit should be investigated and popularized.

Inadequate supply of seeds of improved crop cultivars is a major obstacle to horticultural development in Nigeria and underscores the importance of obtaining information on potential seed yields and cost of production on plots of commercial size. Vegetable production in Nigeria is largely unmechanized and carried out by small farming families in plots less than 0.4 ha.

The process of modernization, urbanization and diversification has led to accelerated change from primitive to relatively sophisticated market garden systems (Grubeen, 1992). Commercial vegetables, often grown more extensively—similar to arable crop production—require 100-250 mandays per ha from establishment to maturity.

Available statistics show that production costs of vegetables (both exotic and indigenous) have been rising and profit margins decreasing due to increased labour costs (although family labour is still utilized), cost of implements, seeds, fertilizers, agrochemicals and packaging materials. Farmers also have to absorb heavy losses due to the high perishability of vegetables and volatile market forces. The size of plots under vegetables is generally small but varies widely in different agroecological zones. In the low country dry zone, the average farm size varies from 0.09 to 0.37 ha. This poses an important production constraint that has to be rapidly alleviated if the economic well-being of farmers is to be improved.

This study examines the resource use efficiency of *Amaranthus*, *Celosia* and *Cochorus* in Oyo state, Nigeria.

MATERIAL AND METHODS

The study area was Ibadan North-west Local Government area of Oyo State, sharing a common boundary with Ibadan North-east Local Government area in northern Ibadan, Ibadan South-west and South-east Local Governments in the south, Ibadan North Local Government to the east and Ido Local Government area to the west.

The area is characterized by two rainy seasons and hence commonly two growing seasons. As an urban centre, there is little land available for agricultural activities. Backyard farming and cultivation along stream banks or the inland valley predominate in the study area. Primary data were obtained through well-structured questionnaires passed out to randomly selected leafy vegetable farmers in the study area and analyzed using descriptive statistics in presenting the responses to certain variables.

RESULTS

Land area allocated to the cultivation of *Amaranthus, Celosia* and *Cochorus* (Table 31.1).

Table 31.1 Land area allocated to vegetables

Land area (ha)	*Frequency*	*Percentage*
Amaranthus		
0.001–0.009	17	22.67
0.01–0.0375	44	58.67
0.04–0.2	14	18.67
Total	75	
Celosia		
Land Area		
0.001–0.009	6	16.23
0.01–0.0375	28	75.68
0.04–0.2	3	8.1
Total	37	
Cochorus		
Land area		
0.001–0.009	8	12.8%
0.01–0.0375	51	72%
0.04–0.2	11	11.42%

Amaranthus: About 58.6% of the respondents place between 0.01 and 0.0375 ha under cultivation of this crop, 22.67%—0.001 to 0.009 ha and 18.67%—0.04 to 0.2 ha.

Celosia: Of the 37 farmers who cultivate this crop in the study area, 75.68% plant between 0.01 and 0.0375 ha, 16.23%—0.001 to 0.009 ha and 8.1%—0.04 to 0.2 ha.

Cochorus: Of the 70 farmers in the study area involved in the production of Cochorus, 72% cultivate between 0.01 and 0.0375 ha, 12.86%—0.001 to 0.009 ha and 11.42%—0.04 to 0.2 ha.

RESOURCE USE EFFICIENCY

Analysis of the productivity of the three leafy vegetables selected for study was carried out. Estimates of the production function coefficients were the direct measures of the input transformation ratios. Four production functions were then fitted into the data, namely Linear, Semilog, Double-log and Exponential for *Amaranthus* (Table 31.2), *Celosia* (Table 31.3) and *Cochorus* (Table 31.4) and a lead function selected for each vegetable.

From a critical look at the four functional forms based on R^2 coefficients and F values and their levels of significance, it can be inferred that double-log is the lead production function that best fits the model.

The lead equation selected has the highest R^2 (0.62), which is significant at 1%; this means that 62% of the explanatory variables (i.e., land, capital, seed and fertilizer) determined the output of *Amaranthus* while 38% (X_{11}, X_{21}, X_{41}) due to error term were not explained. The chosen lead functional form also has the highest F value of 28.55 (sig. at 1%), which means that 1% of the explanatory variables explain the error term of variation in output of *Amaranthus*. Hence an increase in all the explanatory variables in production, except fertilizer, due to their positive beta coefficients would bring about an increase in total output of this crop, while the negative beta coefficient of fertilizer indicates excessive use and any further increase in usage would reduce production of this vegetable.

Table 31.2 Regression analysis of *Amaranthus* output based on four functional forms

Functional form	*Constant term*	*Land* X_{11}	*Capital* X_{21}	*Seed* X_{31}	*Fertilizer* X_{41}	R^2	F
Linear	–538.45 (552.59)	–0.90503 (4.1493)	5.9966 (1.6604) ***	1.5858 (0.58781) ***	1.4427 (1.6830)	0.41	12.29***
Exponential	5.9991 (0.23775)	0.4311E-03 (0.17852E-03)	0.28652E-02*** (0.71436-E03)***	0.73965E-03 (0.25280E 03)	0.91316E-03 (0.72411E-03)	0.47	15.67***
Semilog	–12979 (3316.6)	203.05 (126.87)	2070.4 (548.40)***	624.20 (351.38)*	124.00 (312.76)	0.43	13.04***
Double-log	0.3411E-01 (1.2263)	0.13145*** (0.46910E-01)***	0.85652 (0.20277)***	0.52549 (0.12992)	0.20994-E-01(0.11-564)	0.62	28.55***

***Mean significant at 1% level.
**Mean significant at 5% level.
*Mean significant at 10% level.

Table 31.3 Regression analysis of *Celosia* output based on four functional forms

Functional form	*Constant term*	*Land* X_{11}	*Capital* X_{21}	*Seed* X_{31}	*Fertilizer* X_{41}	R^2	*F*
Linear	–555.05 (342.19)	–2.1347 (0.72226)	0.19639 (0.25749) E-01	0.49179 (0.24133)***	0.55396 (1.3883)	0.72	20.92***
Exponential	6.5002 (0.15411)	0.7531E-03 (0.32528E-03)***	0.86088E-04 (0.11597E-04)***	0.25191-E-03 (0.10869E-03)***	0.35926E-03 (0.62527E-03)	0.72	19.96***
Semilog	–17298 (2490.6)	–2.4385 (0.74602)	1768.2 (247.93)***	628.16 (238.86)***	–131.47 (248.15)	0.71	19.41***
Double-log	–1.5961 (1.0417)	0.86896E-03 (0.31202E-03)***	0.81368 (0.10370)***	0.29316 (0.99904E-01)***	–2.5268E-01 (0.10379)	0.74	22.72***

Table 31.4 Regression analysis of *Cochorus* output based on four functional forms

Functional form	*Constant term*	*Land* X_{11}	*Capital* X_{21}	*Seed* X_{31}	*Fertilizer* X_{41}	R^2	*F*
Linear	47.880 (1541.6)	-8180 (11332)	0.18595 (3.6953)	1.9633 (1.1718)*	5.5587 (3.9075)	0.079	1.39
Exponential	5.8362 (0.39872)	-0.15142 (2.9308)	0.23929E-03 (0.95574E-03)	0.75940E-03 90.30308E-03)	0.48327E-03 (0.10106E-02)	0.23	4.79
Semilog	-5865.3 (124.82)	163.41 (564.40)	-372.46 (1511.6)	1019.5 (803.23)	892.17 (716.55)	0.084	1.50
Double-log	4.6216 (2.8130)	0.39896*** (0.12719)	0.31474 (0.34065)	0.31315* (0.18102)	0.10105 (0.16148)	0.42	11.64***

From a critical look at the four functional forms based on their R^2 coefficient and F values with their levels of significance, it can be inferred that double-log is indeed the lead production function that fits the model. A lead equation with a coefficient of multiple determination (R^2) of 0.74 means that 74% of the total variations in aggregate value of output (Y_{12}) of *Celosia* were attributable to the four explanatory variables—land (X_{12}), capital (X_{22}), seed (X_{14}) and fertilizer (X_{42})—included in the production function analysis. The lead equation is also significant at 1% with an F value of 22.72, which is the highest of the four functional forms.

As mentioned above, a critical look at the four functional forms based on their R^2 coefficient and F values together with their level of significance showed that double-log should be selected as the lead production function best fitting the model. The R^2 (0.42) of the lead equation, the highest of all the functional forms, indicates that 42% of the total variations in aggregate value of output (Y_{13}) of all *Cochorus* were attributable to the four explanatory variables included in production function analysis.

The lead equation is also significant at 1% with the highest F value of 11.64 over the other functional forms. The positive values of the beta coefficient of all the explanatory variables of *Cochorus* means that an increase in land, capital, fertilizer and seeds would bring about an increase in its total output.

References

FAO, United Nations, Rome. 1998. Vegetable Production and Marketing under Arid and Semi-arid Conditions in Tropical Africa. FAO, Via delle Terme di Ceracalla, 00100 Roma, Italie.

Grubeen, G.J.H. 1992. Tropical vegetables and their genetic resources. International Board for Genetic Resources, Rome.

Rice, R.P., Roce, L.W. and Tindal, H.D. 1993. Fruit and Vegetable Production in Warm Climates. Macmillan Education Ltd., London.

32

SOCIO-ECONOMIC SURVEY OF URBAN FRINGE VEGETABLE FARMERS IN ETI-OSA AND SURU-LERE LOCAL GOVERNMENT AREAS OF LAGOS STATE, NIGERIA

M.G. Oludije and L.A. Akinbile
University of Ibadan,
Department of Agricultural Extension and Rural Development,
Ibadan, Nigeria
e-mail: oladele20002001@yahoo.com

INTRODUCTION

Lagos State is not noted for much agricultural production despite the wide expanse of land available in areas such as Epe, Ikorodu, Ibeju Lekki, Eti-Osa, Badagry and some parts of Ikeja. Most of the goods consumed by the inhabitants are therefore mostly brought in from other parts of the country and neighbouring states. This creates a problem of price-fixing since the price of imported crops results in price competition. However, the recent realization of the economic importance of vegetables, especially during the dry season when they become scarce due to a drop in production coupled with increase in demand, has led to their production in fringe, unconventional areas, usually with regular water supply, especially during the dry seasons.

This has prompted unemployed migrants in the state to venture into the production of vegetables year round as a stable source of income and job security.

Vegetables are produced mainly and intensively in the urban fringe of towns and cities of Lagos, owing to high urban demand also the high costs of transportation of vegetables from rural production centres located some distance away from urban markets (Omidiji, 1978). Vegetables are a valuable supplement to the diet of town dwellers. They are also a valuable source of vitamins, protein and minerals, but a cursory survey in tropical and subtropical areas indicated that the average daily consumption of edible greens is about 0.02–0.025 kg (Adetimirin, 1996).

This low consumption can be attributed to both the frequent shortage of vegetables during the dry season in Nigerian markets and their steek prices. Vegetables can be classified into various categories according to the parts consumed as

Leafy vegetables:

- *Amaranthus*, (tete)
- *Celosia* (shoko)
- *Vernonia* (bitter leaf)
- Lettuce

Fruit Vegetables—Tomatoes, okra, egg-plant (Ajayi, 1998).

Root Vegetables—Carrot, Irish potato, etc.

The major concern of this study is to identify those socio-economic factors affecting productivity of fringe vegetable farmers in the area of study.

The importance of vegetables in our daily life cannot be overemphasized in that they are also useful as raw materials in industries and as sources of foreign exchange earnings through exports.

Nigeria is a country blessed with good soils and climate for agricultural purposes. The country also records different varieties of vegetables, both local and exotic ones. Many factors have been shown to influence the type of vegetable production that takes place in a particular area. These include climate, proximity to markets and types of markets, availability of labour, land, capital and other resources, as well as size of holding. The pattern of farming and family tradition are important, but the types and properties of the available soils will have a major influence on the pattern of production and on the types to be grown (Patel, 1980; Olowu, 1996). For improved vegetables, fertilizers are mostly needed to supply plant nutrients and to improve the physical conditions of the soil.

Vegetable growing is one of the most important branches of horticulture in view of the value of its products. About 78% of the total vegetables consumed per capita are eaten fresh, while the remaining 11%

are processed into tomato paste and puree, or dried okra and chillies (Jolomi, 1997). A wide range of annual vegetables are grown in most urban centres of West Africa because of their increased demand, necessitating a high level of intensively cultivated plots in towns and urban fringes.

MATERIAL AND METHODS

The data in this study were collected mainly through a questionnaire survey, supplemented with oral interviews. The questionnaire was designed to elicit information on social, economic and cultural aspects of the selected fringe vegetable farmers and completed on a face-to-face basis, with the responses recorded by the interviewer. However, only 42 of the 51 questionnaires filled in proved useful in the empirical analysis below.

The 42 respondents were divided into a limited number of social groups on the basis of four social, one economic and one cultural variable.

RESULTS AND DISCUSSION

Social Characteristics

1. Gender: All the respondents were men. Women are not directly involved in planting in the study area at these sites; they simply buy vegetables from the men after harvesting. Since all the respondents were men, gender was not used as a social variable in data analysis.
2. Age Range: The vegetable farmers ranged in age from 23 to 46 years. Although they all belonged to the very active population age group, they were divided into three groups: 10 fell in the age group 21–30, 26 in the 31–40 age group and 6 in the 41–50 group. Those in the middle age group (i.e., 31–40) predominated. The labour requirement of the venture may be responsible for this as an artificial supply of water is relied upon to nourish the vegetables since production is mainly during dry seasons.
3. Social Status: All the respondents interviewed were peasant farmers. Membership in cooperative societies was further used as a classificatory criterion. Twenty-eight (28) were cooperative members (of which six were committee members of the local vegetable farmers cooperative society) while 14 were non-members. Respondents belong to cooperatives in order to obtain

loans for procuring the inputs needed for production. This is necessary as the farmers lack needed collateral to obtain formal loans.

4. Educational Status: Twenty-six (26) of the 42 respondents had attended primary or non-formal schools, but none held a first school leaving certificate. The remaining 16 had never attended school. This affects their reception of improved technologies for vegetable production as education has a significant relationship with adoption of new ideas.

Economic Factor

Average income: The weekly income quoted by the respondents varied widely, between ₦ 5000 and 37,000. The wide variation may be due in part to the differential degree to which the respondents were willing to make their income known. It is well known that people do not generally like to disclose such information. The author noted that most respondents were hesitant not only about this aspect of the questionnaire, but that of family size (number of wives and children). Since there was no confidence in these two sets of information, they were dropped in subsequent analysis.

Cultural Factor

Growing Taboos: The respondents were mainly Ilajes, Ijebus, Ghananians and Hausas. Given the anticipated differences between them in terms of social background, there are variations in cultural factors that affect their respective growing habits. This mainly pertains to types of fabricated tools used and the management practices adopted.

Table 32.1 shows that all the socioeconomic characteristics of age (X^2 = 9.01), education (X^2 = 7.16) and marital status (X^2 = 6.02) had a direct effect on farmer production level at the 0.05 level of significance. Older farmers thus had a higher production level than younger ones as did those with a higher SES.

Table 32.2 shows that the problems militating against the business were high cost of fertilizer, pests and diseases, and spoilage resulting from lack of storage facilities and inadequate facilities. Education on optimum combination of inputs was also identified as a problem militating against vegetable production in the study area.

Table 32.1 Chi-square relationship of socioeconomic characteristics with production level

Characteristic	*X^2 Value*	*Df*	*P*	*Decision*
Age	9.01	2	0.04	S
Education	7.16	4	0.01	S
Marital status	6.02	2	0.00	S
Socio-economic status (SES)	7.12	2	0.00	S

S = Significant at 0.05.

Table 32.2 Problems militating against the business (n = 160)

Problems	*Frequency*	*Percentage*
Cost of fertilizer	23	54.8
Pests and diseases	37	88.1
Spoilage resulting from lack of storage facilities	34	80.9
Inadequate credit facilities	40	95.2
Education on optimum combination of inputs	42	100
Multiple response	80	50

CONCLUSIONS

The study revealed that the production practices of the respondents favour market-oriented production aimed at supplying the teeming population and local market. However, there is need for alleviating the problems militating against the vegetable growing farmers in Lagos State so that improved and sustainable production of fringe vegetables can be maintained through education about the optimum combination of inputs for optimum economic outputs.

Fringe farmers make more profit than conventional vegetable farmers who rely on rainfed agriculture since production is generally low in the dry season this is the very season, contrarily, in which income peaks for fringe farmers because the market is algei for their produce. They should therefore be assisted in sustain their efforts, as a means of alleviating individual and communal poverty.

References

Adetimirin, V.O. 1996. Personal Communication on "Vegetable Production". Agronomy Dept., University of Ibadan, Nigeria.

Ajayi, A.O. 1998: Assessment of economic returns of leafy vegetables in Ibadan. B.Sc. Thesis, Dept. of Agric. Economics, Univ. of Ibadan, Nigeria.

Jolomi, E.O. 1997. The contribution of urban fringe farming to Agric. Development of Osodi-Isolo and Ojo Local Government Area of Lagos State.

Olowu, A.I. 1996. A case study of the landuse patterns of urban fringe farmers in Lagos State. H.N.D. project, School of Agric., Lagos Management and Extension Department, pp. 5–6, 15–17.

Omidiji, M.O. 1978. Tropical leaf and fruit vegetables in Nigeria. Vegetables for Hot Humid Tropics, pp. 6–18.

Patel, U. 1980. A socio-economic survey of rural community in Western Nigeria. J. Rural Eco. Dept. 9 (1) 23.

33

COST AND RETURN OF PRODUCTION OF 'UGU' (*TELFAIRIA OCCIDENTALIS*)

L.A. Akinbile and M.G. Oludije
Department of Agricultural Extension and Rural Development
University of Ibadan, Nigeria.

INTRODUCTION

Conventional production of leafy vegetables in Lagos state, Nigeria, comprises such vegetables as *Amaranthus*, *Celosia*, *Cochorus*, okra and tomato. However, 'Ugu', formerly a delicacy of the 'Igbos' in Nigeria, has recently become a 'must have' among all communities in the state due to growing awareness of its nutritional value. *Telfairia occidentalis* contains 27% protein, 53% oil and 20% minerals and vitamins (Akinbode, 1984). It can thus be consumed as a valuable nutritive vegetable and used for several other purposes.

A review of the literature has shown that human consumption of vegetables is far from sufficient in most developing countries. FAO has recommended a daily intake of 200 g vegetables per capita, which comes to 73 kg per annum, to ensure an adequate nutritive diet (Desai, 1992). Only a few countries can meet or exceed this requirement; vegetable consumption in all African countries is far lower. While hard data are scarce for consumption and production of vegetables, micronutrient deficiency consequent to its absence in the diet is better documented. Two billion people, mostly women and children, are deficient in one or more micronutrients (FOS, 1995). The important preventive role of micronutrients in many human diseases and disorders has long been known, since the beginning of the twentieth century. A global initiative to fight micronutrient deficiencies in developing countries was taken in 1992 at FAO/WHO International Nutrition Conference in Rome, because of the remarkably high prevalence of deficiency symptoms. Micronutrients are those substances which the body requires only in small amounts, namely minerals, trace elements and vitamins, in particular iron, vitamin A and iodine, which top the list of micronutrient deficiency when quantitative aspects of nutrition are considered (NIHORT, 1989).

The need to suppress micronutrient deficiency symptoms has reawakened awareness in Nigeria of the importance of vegetable consumption. It has also led to a surge in production of 'Ugu' in the study area, Lagos, to the extent that almost every vegetable farmer now grows it, and it has become available in almost every market. Thus the time has come to assess the profitability of 'Ugu' production in Lagos through a cost and return analysis. This study therefore poses the question: Is it profitable to produce 'Ugu' in Lagos and what are the constraints facing its production? An answer is given below.

MATERIAL AND METHODS

The snowball sampling technique was employed, that is, an ugu farmer would be interviewed, who would mention other farmers, who would mention still others, until the goal of 60 ugu farmers in the study area was achieved. The personal characteristics of each farmer, such as gender, educational level, farming experience and contact with extension services were recorded. Costs of production of ugu and returns from sales were also ascertained at the individual level to measure the profitability of the venture. The results are presented below.

RESULTS AND DISCUSSION

Personal Characteristics of Respondents

The characteristics of the respondents are given in Table 33.1. It can be seen that the variables age, gender, educational level, farming experience and contact with extension services well differentiate the respondents.

Table 33.1 shows that the majority of the respondents range in age from 36 to 50 years, with none above age 50. Thus the respondents are young, or at most middle-aged. The labour requirement for cultivation of ugu, especially in dry-season production, could be the reason since only those capable of rigorous labour venture into ugu culture.

Table 33.1 Personal characteristics of respondents

Characteristics	*Frequency*	*Percentage*
Age Range		
25–30	5	8.3
31–35	14	23.3
36–40	31	51.7
41–45	7	11.7
46–50	3	5.0
Gender		
Male	34	56.7
Female	26	43.3
Educational Level		
No Formal Education	41	68.3
Adult Literacy	11	18.3
Primary	6	10.0
Secondary	2	3.4
Farming experience		
1–2 Years	29	48.3
3–4 Years	16	26.7
5–6 Years	12	20.0
7–8 Years	3	5.0
Contact with extension		
Never	30	50.0
Bimonthly	13	21.7
Monthly	12	20.0
Fortnightly	5	8.3

The Table also shows that though the majority of respondents are males, a substantial percentage are females. This implies that the respondents produce ugu not just for home consumption, but also and especially to generate income.

Most of the respondents had had no formal education, or very little, with only about 10% having completed primary school and 3.4% secondary. Lack of education could well affect the adoption of improved production practices as earlier studies have confirmed the relationship between level of education and the adoption of innovations.

About 70% of the respondents in Naira (the study area) have cultivated 'Ugu' for no more than 5-6 years. This supports the assertion that 'Ugu' was until recently an uncommon vegetable in the study area. Growing awareness of its nutritional value could well have prompted its cultivation.

Lastly, Table 33.1 shows that about 70% of the respondents do not have regular contact with extension services. This tends to affect production since problems on the farm have to be addressed by the respondents without timely assistance from extension agents. Fortification of extension services to farmers is therefore needed.

Production Costs and Returns

Table 33.2 shows that production costs are low for about 60% of the respondents, who spend less than N15,000 per month in cultivation of ugu and about 70% enjoy high returns, i.e. greater than N40,000. Thus the returns of the majority of the respondents outweigh their production costs. Since the supply of ugu in the study area cannot meet the high demand, prices will rise and production of this crop in the area will become even more lucrative.

Table 33.2 Monthly production costs and returns of respondents in Naira (in N)

Production Costs			Returns		
Amount	Freq.	%	Amount	Freq.	%
5,000–10,000	13	21.7	10,000–20,000	2	3.3
10,000–15,000	24	40.0	20,000–30,000	1	1.7
15,000–20,000	17	28.3	30,000–40,000	14	23.3
20,000–25,000	5	8.3	40,001–50,000	12	20.0
Above 25,000	1	1.7	Above 50,000	31	51.7

Gross Margin of Costs and Returns

It can be seen from Table 33.3 that about 40% of the respondents realize a gross margin of between N10,000 and 15,000. None have a deficit gross margin nor did any produce at a loss. In fact, more than 20% of the

Table 33.3 Gross margin of costs and returns from production of ugu in Naira as reported by study respondents

Gross Margin	*Frequency*	*Percentage*
5,000–10,000	2	3.3
10,001–15,000	24	40.0
15,001–20,000	3	5.0
20,001–25,000	16	26.7
25,001–30,000	1	1.7
Above 30,000	14	23.3

Gross margin = Total revenue minus Total variable cost (GM = TR – TVC)

respondents produce with a gross margin above N30,000. This implies that ugu production alone lifts some people above the poverty line, as indicated by the United Nations (1998) multinational comparison of poverty delineation of US $1.00 (Dianam, 2000). Thus, as already mentioned, ugu constitutes a lucrative business in Naira. The respondents major complaint was lack of wetland to ensure year-round production.

Personal Characteristics of Respondents that Enhance Production

Table 33.4 reveals the significant relationship of level of education, contact with extension services and farming experience with level of farmer ugu production. Gender and age showed no such relationship.

Table 33.4 Personal characteristics of respondents that enhance production

Variable	x^2 *Value*	*r-value*	*df*	*p-value*
Gender	2.07	-	2	0.09
Educational level	3.68	-	2	0.02*
Contact with extension	1.53	-	4	0.01*
Age	-	0.93	58	0.11
Farming experience	-	0.59	58	0.00*

*Significant at 0.05 level.

More educated farmers had a higher level of production than the less educated. Respondents showing greater contact with extension services likewise showed a higher level of production than farmers with less contact, as did those with longer experience in ugu cultivation.

Adult literacy programmes should therefore be organized for ugu farmers as this would encourage their adoption of improved practices and thereby elevate the level of production.

Extension services should also be urged upon the less educated. Obviously, farming experience can only be gained over time. All of these put in place will aid the global effort at alleviating poverty, especially among ugu farmers.

CONCLUSION

The study reveals that education, contact with extension services and farming experience are significant factors in the level of ugu production attained by respondents. The first two can be ameliorated through adult literacy programmes and better provision of extension services and the last over time.

Despite shortfalls in education, advice from extension services and experience in farming, ugu production showed a high gross margin level in the study area. Without doubt, enhancement of the former would augment the latter even more. Furthermore, the vegetable could become more readily available to play a role in counteracting micronutrient deficiencies not just locally, but throughout the state of Lagos, perhaps even all of Nigeria.

References

Akinbode, I.A. 1984. Vegetable production in Nigeria. Nigerian J. Agron. 41: 14-19.

Desai, M. 1992. Population and poverty in Africa. African Develop. Rev., pp. 63-78.

Dianam, D. 2000. Again, the challenge of poverty alleviation. The Guardian Newspaper, February 2000, pp. 19-20.

FOS (Federal office of Statistics). 1995. Statistics Year Book. FOS, Nigeria, pp. 42-47.

NIHORT (Nigerian Institute of Horticultural Research). 1989. Guide to production of vegetables. NIHORT Package Bull., pp. 12-15.

34

INFORMATION SOURCES UTILIZED BY FARMERS FOR VEGETABLE PRODUCTION

O.I. Oladele
University of Ibadan, Department of Agricultural Extension and Rural Development, Nigeria.
e-mail: oladele20002001@yahoo.com

INTRODUCTION

Information sources are important and fundamental stimuli to individual farmers in the adoption of improved farming practices. According to Williams (1969), there are two reasons why these sources of information used by farmers should be known: first, knowing these sources will reduce the cost of extension services and secondly, assist extension agencies in deployment of staff.

Several sources of agricultural information are available to farmers in addition to extension services, namely: public organizations, private organizations, non-governmental organizations (NGOs) and community-based organizations (CBOs).

Lionberger (1960) referred to sources of agricultural information as people and agencies from whom farm information is sought. He grouped these sources as follows:

1) institutional, e.g. extension workers of public organizations, private organizations, NGOs and CBOs;

2) non-institutional, e.g. neighbours and friends, village leaders etc.

Williams and colleagues (1984) defined channels or extension methods as the avenues or media used for passing messages to and from the audience or receiver. Rogers (1973) reported that information sources to farmers could be personal, i.e., face-to-face, or impersonal, i.e., dissemination of information through a mass medium. Williams and Williams (1971) found that extension agents were used by a wide spectrum of farmers as a source of information on improved farm practices and that farmers active in several organizations rarely, if ever, used the radio as an information source.

Bogunjoko (1983) carried out a study in Giwa district, Kaduna state, on sources of improved farm practices. He reported that 92.5% of the farmers interviewed indicated the radio as their source of information, 5% fellow farmers and 4% family or relatives; only 3.5% reported using agricultural agencies.

The importance of disseminating cultural information to farmers led to the constitution of extension services in the Agricultural Development Programme (ADP) and Agricultural Extension Research Liaison Services (AERLS) of research institutes. At present, extension services involve research and input suppliers in a liaison that is both formal and informal to provide information to the end users. Such is the Research Extension Farmers Input Linkage Model (REFIL).

This study analyzes the types of information sources utilized by vegetable farmers in south-western Nigeria.

MATERIAL AND METHODS

The study area comprised seven states, namely Delta, Edo, Lagos, Ogun, Ondo, Ekiti and Oyo and involved the following agricultural research institutes: Institute of Agricultural Research Training (IAR & T) Ibadan, National Institute for Horticultural Research (NIHORT) Ibadan, Cocoa Research Institute of Nigeria (CRIN) Ibadan, Rubber Research Institute of Nigeria (RPIN) Benin and Forestry Research Institute of Nigeria (FRIN) Ibadan. Other consulted institutes were: Nigeria Institute for Oil Palm Research (NIFOR) Benin, National Institute of Oceanography and Marine Research (NIOMR) Lagos, Nigeria Stored Products Research Institute (NSPRI) Ibadan, Federal Institute of Industrial Research (FIIRO) Ibadan, National Agricultural Research and Liaison Services (NAERLS) Ibadan and National Cereal Research Institute (NCRI) Ibadan.

RESULTS AND DISCUSSION

Personal Characteristics of Farmers

Most of the farmers participating in the study were males (81.5%), as shown in Table 34.1. This suggests that farming is still male dominated, notwithstanding the description of the decade past as the 'age of women-farming renaissance' (Egunjobi, 1991). About 71% of the farmers were married and the family size of 74% ranged from 4 to 9. Marriage and a large family size are important in the provision of additional farm labour.

Table 34.1 Personal characteristics of farmers

	Frequency	*Percentage*	*Mean*	*SD*	*SE*
AGE					
<30	22	11	42.9900	12.4598	0.8810
30–40	76	38			
41–50	61	30.5			
51–60	27	13.5			
>60	14	7			
GENDER					
Male	163	81.5	1.8100	0.4414	3.12E-02
Female	34	17			
EDUCATIONAL LEVEL					
No response	10	25.0	2.0900	1.0853	7.674E-02
Not educated	60	30.0			
Primary	50	25.0			
Secondary	63	31.5			
Tertiary	16	8.0			
Others	1	05			
INCOME (in N)					
<10,000	6	3	209,769.25	370,138.72	26,172.7598
10,000–20,000	47	23.5			
20,001–50,000	65	32.5			
>50,000	82	41			

About 68% of the farmers had 2 to 4 dependents, indicating parental obligation to extended families. More Muslims (46.5%) are engaged in farming activities in the area studied than Christians (37.5%). Of the total farmers interviewed, 32% had had some formal education versus 30% uneducated.

The estimated annual income of 41% of the farmers is above N50,000, while about 57% engage in other income-generating activities. Table 34.1 further shows that 38% of the farmers cultivate about 2 ha land.

It can be further seen in this Table that most (57%) of the farmers produce crops. Those belonging to one organization or another constitute 63%.

Organizations Contacted by Farmers

The organizations contacted by farmers, together with mean and standard error, are presented in Table 34.2. Of the public organizations, NAERLS (8.5%), NCRI (7.5%), NSS (6.5%) and NSPRI (8.5%) were the most frequently contacted by farmers. ADOM (Nig.) Enterprises (6.5%) was the private organization most often consulted. All the NGOs, namely FADU (7.0%), DADP (8.5%) and COWAN (8.5%) as well as the CBOs, i.e., cooperative societies (6.5%) and ACOSFA (8.0%) recorded a relatively high farmer contact. In general, the level of contact between farmers and organizations was lower.

Table 34.2 Organizations contacted by farmers

Organization	*Frequency*	*Seldom*	*Very Rare*	*No Response*	*Mean*	*SD*	*SE*
ADP	1(0.5)	91(45.5)	75(37.5)	32(16.0)	1.4150	1.7021	0.1204
NIHORT	2(1.0)	83(41.5)	74(37.0)	41(20.5)	1.2300	0.7811	5.523E-02
NSS	13(6.5)	89(44.5)	38(19.0)	60(30.0)	1.2750	0.9665	6.834E-02

SUMMARY

The major information sources utilized by vegetable farmers in the process of agricultural production have been indicated. These sources are highly diversified and sufficiently numerous to reach the majority of farmers, most of whom live in rural areas where other information facilities may be wanting.

References

Bogunjoko, I.O. 1983. Sources of information of improved farm practices. A study of farmers in Giwa District of Kaduna State. Nigeria J. Agric. Exten., 1 (2).

Egunjobi, O.A. 1991. Harnessing the woman factor in the Nigeria food security drive. The role of the University, pp. 41-49. In: T.A. Olowu and J.A. Akinwumi (eds.). Proc. Natl. Conf. Ibadan Socio-econ. Group July 7-18, Univ. Ibadan, Nigeria.

Lionberger, H.F. 1960. Adoption of New Ideas and Farm Practices. Iowa State University Press, Ames, IA.

Oladele, O.I. 1999. Analysis of the Institutional Research-Extension-Farmers Linkage system in south-western Nigeria. Ph.D. thesis, Dept. Agric. Extension Services, Univ. Ibadan, Nigeria.

Otuonye, J.C. 2000. Activities of private agencies involved in extension services in south-western Nigeria. M.Sc. thesis, Dept. Agric. Extension and Rural Development, Univ. Ibadan, Nigeria.

Rogers, E.M. 1973. Communication Strategies for Family Planning. Free Press, NY USA.

Williams, S.T.K. 1969. Sources of information on improved farm practices in selected areas of Western Nigeria. Bull. Rural Econ. Sociol., 1 (2).

Williams, S.K.T. and Williams, C.E. 1971. Farmer contact with agricultural extension services in western states of Nigeria. Bull. Rural Econ. Sociol., vol. 6.

Williams, S.K.T., Williams, C.E. and Fenley, J.E. 1984. A Manual for Agricultural Extension Workers in Nigeria. Les Shyrens Publishers, Ibadan, Nigeria.

35

SPATIAL DISTRIBUTION OF ON-FARM ADAPTIVE TECHNOLOGIES FOR VEGETABLE PRODUCTION

O.I. Oladele
Department of Agricultural and Rural Development,
University of Ibadan, Nigeria
e-mail: oladele20002001@yahoo.com

INTRODUCTION

Nigeria's total land area is estimated to be approximately 91.08 million hectares, extending from latitude 4°14′ N to 13°48′ N and longitude 2°42′ E to 14°40′ E. The land-use pattern shows that about 57% of the land is under crop and pasture. The diversity of the plant cover shows transition from mangrove swamps to the rain-forest zone through the derived savanna belt to the Guinea Savanna and Sudan Savanna, terminating at the Sahel Savanna. Given this diversity of plant cover, the country was divided into five agricultural zones, reflecting the varied agroecological environment of Nigeria, based on the premise that such division would strengthen farming research and research extension links by allocating responsibility for each zone to a specific research institute within it.

Vegetable production in Nigeria is spatially distributed along the country's ecological setting, the five zones differing in intensity of cultivation. Gibberd and Gibberd (1979) noted that the vegetable varieties that can be grown in the tropics and the method of cultivation employed are largely determined by altitude and proximity to the coast. At very high altitudes (5000 feet or more), conditions approximate more closely those that obtain in temperate zones. In the rain-forest, vegetable production is difficult while in the savannas the best results are obtained in the dry season.

Recognition of the above described zones has naturally influenced the distribution of on-farm adaptive technologies (OFAR) for vegetable production. OFAR is an integral part of the Research-Extension-Farmer Input Linkage System in Nigeria, which allows researchers to adjust general recommendations derived during on-station assays to different farm situations. Not only researchers, but extension agencies and farmers also usually manage OFAR.

AGRICULTURAL ZONES IN NIGERIA

The north-west zone lies between latitudes 9° N and 14° N and has an annual mean rainfall ranging from 500-1200 mm. Vegetation is typically savanna and major soil types Lixisol, Regosol, Terasol, hydromorphic Fluvisol and Gleysol. Vegetables are produced from valley bottom-land. This zone contributes 79% of the onion, 48% tomato, 42% pepper and 16% leafy vegetables produced in the nation (FMANR, 1997).

The north-east zone lies between latitudes 8° N and 14° N, with an average annual rainfall of 500-1000 mm. Vegetation is of the Sahelian savanna type. Abundant quantities of tomato, onion, pepper, carrot and miscellaneous vegetables are produced. OFAR technologies disseminated in the area include: use of local and improved vegetable varieties of cabbage, sweet pepper, okra, spinach, onion—Texa and Grano, garlic (Ex-kota variety), Telferia NR-8083, tomato—Romar VFN; dry season vegetable production; vegetable crop planting on ridges; introduction of tolerant tomato cultivar trials; and bulk storage of fresh vegetables employing an evaporating cooling system (FMANR, 1997).

The central zone is situated between latitudes 6°30′ and 11°20′ N, with extensive plains at an altitude less that 150 m above sea level. The mean annual rainfall for the zone ranges from 1000-2000 mm and vegetation comprises grasses and palms, shrubs and grasses as well as trees. Vegetable production is not commonly practised in the zone. However, OFAR technologies disseminated in the area include: vegetable production using local and improved varieties and dry season vegetable production (FMANR, 1997).

The south-west zone lies between 5° and 9° N and has a mean annual rainfall of 1200-2600 mm. Rain-forest and derived savanna predominate. Nevertheless, the zone produces large quantities of melon, okra, tomato and pepper, and boasts the highest intensity of vegetable production in the nation. OFAR technologies in this area include: vegetable production using local and improved varieties, sweet pepper, okra, spinach, onions—Texa and grano, Garlic (Ex-kota variety) and tomatoes—Romar VFN; dry season vegetable production; vegetable cropping on ridges; and tolerant tomato cultivar trials.

The south-east zone lies between 4° and 7° N and is predominantly plains less than 700 m above sea level. Mean annual rainfall varies from 1500-3000 mm and humid forests predominate. The zone reputedly produces 49% of the leafy vegetables in the nation. OFAR technologies disseminated in the area include: vegetable production using local and improved varieties such as Telferia NR-8083, vegetable cropping on ridges and introduction of tolerant tomato cultivar trials (NAERLS/APMEU, 1997).

CONCLUSION

The zonal nature of Nigeria is a major factor in vegetable production in the country, but social, cultural and economic factors might well be responsible.

References

FMANR. 1997. Nigeria: National Agricultural Research Strategy, Plan 1996-2010, pp. 41-56. Bukar Shaib Adamu Aliyu and J. S. Bakshi (eds.). Department of Agricultural Scenes, Abuja.

Gibberd, A.V. and Gibberd, V.L. 1979. A Gardening Notebook for the Tropics. Longman Group Ltd., London, p. 55.

NAERLS/APMEU. 1997. Prospects and Problems of the 1997 Cropping Season. NAERLS: Ahamadu Bellow University, Zaria, 81 pp.

36

SOCIO-ECONOMIC FACTORS AFFECTING LOCUST BEAN MARKETING

O.A. Adekunle and O.I. Oladele
Department of Agricultural Extension and Rural Development,
University of Ibadan, Nigeria.
e-mail: oladele20002001@yahoo.com

INTRODUCTION

African locust bean (*Parkia filicoidea*) is a tree legume found throughout West Africa, South-east Asia and tropical South America. It is a medium-size tree found in open savanna woodland where it may be the only tree species at places, characterized by fruits that have elongated pods, 5–11 inches long, occurring in clusters. The immature fruit is green, turning a brown as it matures. The mature pod consists of a husk which encloses a powdery yellow pulp in which dark brown seeds are embedded.

The seeds constitute about 25.04% of the weight of the pod. The whole fruit contains on a dry matter basis 12.7% protein, 6.8% fat, 6.2% ash and 18% fibre. The mature seed contains about 30% digestible carbohydrate (starch and sugars), 30% protein, 20% fat, 9% crude fibre and 5% minerals. Not only is the fat content higher than in most grain legumes, it is made up predominantly of unsaturated fatty acid (54% total fatty acids). The unsaturated fatty acids are linoleic acid (42% of total), oleic acid (8.8%) and palmistoleic acid (2.7%). The saturated fatty acids which contribute 46% of the total fatty acids, comprise palmitic acid (31%), stearic acid (7.7%), arachidic acid (4.2%) and behenic acid

(31%). The powdery yellow pulp contains more carbohydrates than the seed, the carbohydrates being primarily reducing sugars (19%), non-reducing sugars (9%) and other complex carbohydrates (36%).

Almost every part of the tree serves a useful purpose in the household. The leaves are rich in nitrogen and ash, serving as an important source of green manure for homestead agriculture. The whole pod provides valuable fodder for animals, particularly in the dry season when animal feed shortages are common. The yellow pulp may be used as an ingredient in the preparation of various stews, soups and sauces for the consumption of rice, cereals and grain legumes. The pulp may also be pressed into cakes and preserved for later use or used in the preparation of various indigenous drinks. The dark brown seeds are the most valuable part of the crop. They are not eaten raw, probably because of their high content of indigestible components and antinutritional factors (Odunfa, 1985).

The seeds are fermented into a very important flavouring condiment in West Africa, which is used in much the same way as bouillion cubes in the western world. In Northern Togo, some of the local people grind the seeds to a fine meal and steep it in water to make a refreshing gruel.

Fermentation significantly improves the nutritional value of the locust bean and makes it one of the important sources of dietary protein in those areas of the world where it constitutes a food ingredient in the human diet. Fermented locust bean cakes are eaten in various countries as a meat substitute, especially during periods of famine and starvation, which are endemic to many countries in Africa and Asia. It may be combined with pepper and salt and turned into sauces for consumption of cereals, cassava, yam and other roots and tubers. In northern Nigeria, it has been estimated that the locust bean ranks third in order of importance as a source of dietary protein intake, after guinea corn and millet. In general, it contributes 1.5% of the caloric intake and 5% of the protein intake of the Hausa people of northern Nigeria (Simon, 1986). Locust bean seed has a very hard, water-impermeable testa and therefore prolonged cooking is necessary during processing to soften the seed coat prior to dehulling. According to Odunfa (1985), presoaking for 12 hours and cooking for 30 minutes in a pressure cooker suffices to soften the seed, but removal of the seed-coat is more difficult. Pressure cooking for 90 minutes without presoaking facilitates removal of the seed coat while sufficiently softening the seed. Seeds may be dehulled manually or by machine. Machine dehulling before cooking is practised as an easier alternative since the hydrophobic seed-coat which prolongs cooking is eliminated.

After the initial cooking regime, the dehulled seeds are cooked for an additional 30 minutes and gently pounded to further soften them. Spoonfuls of the pounded seeds are wrapped in leaves (preferably plantain leaves) and stored in various-size containers for fermentation. Piling of the wrapped seeds in storage basket causes the temperature of storage to rise to a level adequate for initiation of fermentation.

Fermentation occurs in 48-27 hours at an ambient temperature of 25-35°C. Fermentation depends on chance inoculation from the environment by organisms identified by Odunfa (1985) as belonging to *Bacillus subtilis* coagulase negative *Staphylococcus, Bacillus licheniformis* and *Staphylococcus saprophyticus. Enterobacter aerogenes* and *Coryneformis* bacteria have also been identified (Kuboye, 1982).

Chance inoculation with these various microorganisms make the quality and flavour of the final product inconsistent. The bacteria produce viscous strands of mucilaginous material which link the beans together and produce a strong ammoniacal odour. During fermentation, there is extensive hydrolysis of protein to amino acids and of oligosaccharides to simpler sugars. Hydrolysis of phytates and oxalates during fermentation also releases minerals, rendering them more available, especially calcium and phosphorus. Protein and fat content of the fermented product is much higher than in the locust bean seed, at 40% and 35% respectively in the fermented cakes. Riboflavin and thiamine may increase threefold and sixfold respectively following fermentation. However, fermentation causes a one-third decrease in niacin content. Fermented seeds are partially dried in the sun during which they darken in colour.

When partially dried, the beans are moulded into balls or shaped into flat cakes of about 5 cm in diameter and further dried to improve their shelf-life and make them stable for at least 6 months (Campbell-Platt, 1986).

The level of awareness of the nutritional value of locust beans is low. Most people complain about the odour of the fermented seeds but were they educated about their nutritional value, buyer interest would rise. It is also thought that some constraints in marketing of locust beans exist and these should be investigated. Societal values account for the large number of women interested in marketing locust beans, which influences the level of planting of the commodity. Locust beans are not readily available in the market, especially during the dry season. If preservation methods or equipment could be made available, scarcity would be minimal and that would boost the economic status of the sellers.

The general objective of this study was to investigate the social and economic characteristics of the women selling locust beans in Oyo state in Nigeria.

MATERIAL AND METHODS

The study area was Atiba Local Government Area of Oyo state. It is one of thirty-three local government areas in Oyo and is bounded in the north by the Oyo Local Government Area, in the south by Akinyele Local Government Area, in the west by Iwo Local Government Area of Osun state and in the east by Iseyin Local Government Area. The people living in the Atiba Local Government Area are predominantly Yorubas, while other tribes such as the Ibo, Hausa, Nupe etc. constitute a minority. The predominant language is Yoruba. The inhabitants of Atiba are predominantly farmers but also engaged in secondary occupations such as barbing, bricklaying, craftwork, mat-weaving, tailoring, trading etc.

The population studied consisted of (all) women selling locust beans in various markets in the Atiba local government area. The market functions every five days and farmers from various areas within and outside the local government bring their farm produce for sale.

The total number of women marketing locust beans in Atiba 187, among which purposive sampling was done to select 90 respondents.

RESULTS AND DISCUSSION

Demographic Characteristics of Respondents

The characteristics discussed include age, marital status, level of education, religion, occupation, income and standard of living. A total of 90 respondents were interviewed, all of them females.

As shown in Table 36.1, 6.7% of the respondents fall in the age group 26–35 years. The reason for the small number of people in this age range might be due to migration of youths from rural areas to urban areas in search of jobs and furtherance of their education.

Respondents of 36–55 years old formed the largest proportion of the population studied (85.6%). These able-bodied people are still very active in commercial endeavours. Seven per cent of the respondents were age 56–65 years, women less able to cope with commercial activities, which accounts for their small number.

According to the Table 36.1, 85.8% of the respondents had between five and nine people in their households. This observation supports the view that there is a high dependency ratio in the area of study. Ten

Table 36.1 Demographic characteristics of respondents

Characteristic	*Frequency*	*Percentage*
Age (year)		
26–35	6	6.7
36–45	36	40.1
46–55	61	45.5
56–65	7	7.7
Total	90	100.0
Family Size		
1–4 people	10	10.1
5–9 people	76	85.8
10–14 people	4	4.1
Total	90	100.0
Level of Education		
Tertiary	5	5.6
Secondary	6	6.6
Primary	54	60.0
Adult Education	1	1.1
No formal education	24	26.7
Total	90	100.0
Income		
Less than N10,000	69	76.6
N10,000 – N18,000	21	23.4
N18,000 and above	–	–
	90	100.0

percent had between one and four people in their households, peculiar to women of recent marriage.

The level of education of the respondents was very low. Fifty-four had primary school education, 24 no formal education whatsoever, 5 left the polytechnic with an Ordinary National Degree Certificate and had secondary education. Only one woman had attended adult education classes and did so outside the study area. Given the aforesaid variation in level of education, the women did not recognize the need for group action, each relying on individual efforts in getting things done.

The standard of living of the respondents was considerably low. Items such as stove, iron, fan, mattress, television, radio, fridge etc were listed for the respondents to indicate which they possessed and which they had bought with their personal money. The number of items claimed to be possessed were calculated by the current price of 'second-hand' items in the market and this parameter used as the indicator of their standard of living.

Table 36.2 Factors motivating respondents to continue selling locust beans

Factors	*Frequency*	*Percentage*
Nutritional value	17	18.9
Profit	62	68.9
Mother's business	11	12.2
Total	90	100.0
Quantity		
Bags	59	65.6
Tins	25	27.8
Kongos	6	6.6
Total	90	100.0
Number		
1–2 bags	45	50.0
2–3 bags	23	25.6
4–5 bags	22	24.4
Total	90	100.0
Sources of information		
Friends	64	71.1
Farmers	16	17.8
Agric. Programmes on radio	10	11.1
Total	90	100.0

Table 36.2 shows that 69% of the respondents noted that they are motivated to continue selling locust beans because of its profitability, 19% claimed motivation because of nutritional value while 12% indicated they had inherited the business from their mothers. This conforms with the fact that most commercial ventures are embarked upon for the purpose of making a profit, inclusive of agricultural products.

Sixty-six per cent of the respondents bought locust beans in bags for retail purpose because economically they could afford to do so. Further they believed their profit would be greater from buying in bulk and they would be less affected by periodic product scarcity. Twenty-eight per cent bought locust beans in tins while only 7% bought in kongos. The latter buyers made far less profit than the women who bought in bags.

Fifty per cent of the respondents sold between 1 and 2 bags per month. These are the women who bought locust beans in 'kongos'. One bag is equivalent to 30 'kongos' or to 3 tins. A 'kongo' sold for N110 at the time of this investigation and a bag rounded off at N3300 after sale. About 26% sold 2–3 bags while 24% sold between 4–5 bags. Women who sold less quantities were more involved in processing 'iru' (the fermented beans) to supplement their income.

Effective marketing of agricultural products according to Anthonio (1984) depends on accuracy of market information and channels of distribution for farm produce. About 71% of the respondents obtained information on the availability of locust beans from friends who were also traders. This shows that information channels here are not very effective since the information passed might sometimes be in accurate. Only 11% of the respondents listened to agricultural programmes on the radio to obtain information; some women even delegated their wards to listen to the radio and relate the information later.

Testing of Hypotheses

As evident from Table 36.3 there is no significant relationship between the socio-economic status and age of the respondents and marketing of locust beans. As the correlation between family size and marketing is negative, a null hypothesis is accepted, namely there is no significant relationship between family size and marketing.

As for the relationship between social infrastructure, e.g. road network and marketability of locust beans (HO_2), being non-significant, it can be seen from Table 36.4 that chi-square computed is greater than chi-square tabulated; therefore the null hypothesis is rejected. On the contrary, there is a significant relationship between social infrastructure, e.g. road network and marketing of locust bean. This supports the work of Kohl (1987) who said that a 'good transportation system cannot be overlooked when discussing marketing efficiency, because middle men have to travel to various markets for supply before transporting them to

Table 36.3 Relationship between selected demographic characteristics of respondents and marketing

Variables	*r*	*p*	*Decision*	*Remark*
Socio-economic status	0.12	0.24	N.S	Accept H0
Age	0.07	0.53	N.S	Accept H0
Family size	– 0.01	0.04	N.S	Accept H0

Table 36.4 Summation of Chi-square (X^2) analyses

Road network	Computed X^2 12.13	Tabulated X^2 5.99	df = 2 at 0.05
Credit facilities	Computed X^2 17.07	Tabulated X^2 5.99	df = 2 at 0.05 level
Sources of information	Computed X^2 3.15	Tabulated X^2 7.82	df = 2 at 0.05 level.

a central location'. These movements have to be facilitated by a good road network for the price of agricultural products to be affordable.

Hypothesis: There is no significant relationship between level of accessibility to credit facilities and marketing of locust beans (HO_3).

Chi-square computed is greater than chi-square tabulated; therefore the null hypothesis is rejected and the alternative which states that the relationship between level of accessibility to credit facilities and marketing of locust beans is significant is accepted. Credit facilities are very important in promoting market efficiency and meaningful and economic contributions.

Hypothesis: There is no significant relationship between sources of information available to the respondents and motivation to continue selling locust beans (HO_4).

Computed chi-square is less than chi-square tabulated; therefore the null hypothesis is accepted which states that there is no significant relationship between sources of information available to the respondents and motivation to continue selling locust beans. Respondents noted that they are motivated by the profitability of locust beans.

References

Anthonio, Q.B. 1984. Marketing development in Nigeria. A review relevant to Government Policies.

Campbell - Platt, G. 1986. African Locust Bean *Parkia* spp. and its West African fermented food product, dawadawa. Food Nutrition, pp. 123–132.

Kohl, R. 1987. Marketing of Agricultural Products. The Macmillan Company, NY.

Kuboye, M.A. 1982. Traditional fermented foods and beverages of Nigeria. Provisional report no. 20. Proc. IFS Workshop on development of indigenous fermented foods and food technology in Africa, Dauala, Cameroon.

Odunfa, S.A. 1985. Micro-organism associated with fermented African locust bean during 'iru' preparation. J. Plant Foods, pp. 245–250.

Simon, E.B. 1986. Calorie and protein intake in 3 villages, May, 1970–July, 1982. Simaru miscellaneous paper No. 55. Institute of Agricultural research, A.B.U Zaria, Nigeria.

37

ECONOMICS OF PRODUCTION OF *COCHORUS, CELOSIA* AND *AMARANTHUS*

K.O. Adenegan, A.A. Adegeye and A.B. Odusina
University of Ibadan,
Department of Agricultural Economics,
Nigeria
e-mail: oladele20002001@yahoo.com

INTRODUCTION

Modernization of horticulture has led to an enormous decrease in the number of vegetable species once consumed by our ancestors. It has also narrowed the genetic base of the small number still extant of these species by replacing the local, unselected cultivars with a few highly selected, high-yielding and high-quality cultivars.

Vegetables constitute one of the fastest growing agricultural sectors in both West and Central Africa, driven by rising urban demand. For example, tomato production in Nigeria, Côte d'Ivoire and Ghana more than doubled in the ten-year period 1987-1997, with an estimated annual retail market value of more than US $ 640 million (Rice et al., 1993).

The culinary variations produced from vegetables by various social, cultural and ethnic groups is ample testimony to their importance in providing palatable and nutritionally valuable meals for a family.

Farmer awareness of the economic benefits to be derived from vegetable production is evidenced in judicious timing of establishment of certain crops and application of new technology to increase yields and better quality vegetables as demanded by the market. Without doubt vegetables are a significant factor in the Nigerian economy given the number of farmers engaged in their production and marketing.

Nonetheless several production constraints do hinder realization of the full benefits of vegetables as an important supplementary source of nutrition, mainly development and availability of appropriate production technologies for increasing yields, modern and inexpensive methods of packaging, processing and marketing to meet not only domestic needs, but also exports in a highly competitive market.

Studies during the 1990s have shown that for fruit crops in areas of low desert climates, peak water use can be as high as 11.4 mm day^{-1} and for vegetable crops, 10.16 mm day^{-1} Grubeen (1992) pointed out that since most leafy crops wilt and deteriorate rapidly within a few hours of harvesting, they are mainly grown for domestic consumption with just a few commonly available in the market.

The importance of vegetables depends on the food habits and preferences of the population and local climate. Leafy vegetables consumed in lowland humid areas are generally different from those eaten in higher altitudes.

FAO (1988) revealed an increasing popular awareness of the value of leafy vegetables in a balanced diet, especially in areas of animal protein deficiency. In addition to their iron content, leafy vegetables contain significant amounts of betacarotene and ascorbic acids (vitamin C), protein, minerals (in particular, calcium) and carbohydrates.

The economics of producing just three highly nutritious leafy vegetables—Cochorus, Celosia and Amaranthus in Oyo state, Nigeria—is analysed in this article.

MATERIAL AND METHODS

The study area was Ibadan North-west Local Government area in Oyo state, sharing a common boundary with Ibadan North-east Local Government area in the north, Ibadan South-west and Ibadan South-east Local Government areas in the south, Ibadan North Local Government in the east and Ido Local Government area in the west.

This area is characterized by a bimodal rainfall and thus two growing seasons are recognized. As an urban centre, however, little land is available for agriculture. Backyard farming and cultivation along stream banks or in the inland valley predominate. Primary data were collected for this study through well-structured questionnaires distributed to randomly selected leafy vegetable farmers in the study zone and the data analysed using descriptive statistics for presentation of responses to predetermined parameters.

RESULTS AND DISCUSSION

Farmer-cultivation of Amaranthus, Celosia and Cochorus Ratio

It can be seen from the Table 37.1 that every farmer in the study area cultivates Amaranthus. They consider this vegetable highly marketable (i.e., desired by many consumers), with low gestation period (i.e., takes little time to ready for market and easier to get planting material for). Of the 75 farmers interviewed, 70 (93.33%) also cultivated Cochorus because it, too, commands a high market price, particularly due to its great importance in ceremonial (social) activities.

Only 37 (49.33%) grew Celosia because planting material (seeds) are not so readily available, gestation period is longer (i.e., readying for the market takes longer than the other two leafy vegetables) and the demand is lower.

Table 37.1 Farmer-cultivation of selected leafy vegetable ratio

Vegetable	*Frequency*	*Percentage*
Amaranthus	75	100
Celosia	70	93.33
Cochorus	37	49.33

Cost Analysis

The fixed cost (Table 37.2) incurred was lowest for Celosia at around N11,830.89 because few farmers invested in its production due to economic reasons: high cost of seeds, difficulty in obtaining them and long gestation period.

Profits (returns) were highest for Cochorus at N1,030,026.6 as it sold at higher market prices than either Amaranthus or Celosia, with net returns of N972,904.53 and N379,033.7 respectively.

Table 37.2 Cost analysis of production per hectare in the study area

Selected leafy vegetable	*Output (tons)*	*TFC (N)*	*TVC (N)*	*TC (N)*	*AVC (N)*	*ATC (N)*	*AFC (N)*
Amaranthus	215.73	14330.89	91414.58	105745.47	721.22	834.28	113.06
Celosia	86.404	11830.89	41155.41	52986.3	448.95	578.01	129.10
Cochorus	134.87796	14180.89	79775.56	93956.45	598.35	704.71	106.36

CONCLUSION

Considering the economic efficiency of each of the three leafy vegetables selected for study, it may be inferred that Cochorus is the most efficient at 10.96, followed by Amaranthus at 9.2 and Celosia coming last at 7.15.

References

FAO, United Nations, Rome. 1988. Vegetable Production and Marketing under Arid and Semi-arid Conditions in Tropical Africa. FAO, via delle termedi ceracalla, 00100 Roma, Italie.

Grubeen, G.J.H. 1992. Tropical Vegetables and Their Genetic Resources. International Board for Genetic Resources, Rome.

Rice, R.P., Roce, L.W. and Tindal, H.D. 1993. Fruit and Vegetable Production in Warm Climates. Macmillan Education Ltd., London.

38

DETERMINATION OF SATURATED HYDRAULIC CONDUCTIVITY OF SOILS OF THE KUBANI FLOODPLAINS IN ZARIA, NIGERIA, USING SHALLOW-WELL PUMP-OUT TESTS

A.A. Ramalan
Ahmadu Bello University,
Agricultural Engineering Department,
P.M.B. 1044, Zaria, Nigeria
e-mail: ramalan.aliyu@skannet.com

INTRODUCTION

Soil hydraulic conductivity, popularly described as soil permeability among soil scientists, is one of the major soil physical properties for solving problems in foundation engineering and agricultural land drainage. Problems encountered in these areas include highway and airport drainage, uplift pressure beneath concrete dams, seepage through dams, failure of retaining walls, etc. Those pertaining to agriculture include surface flooding, high water table, man-induced salinity, etc. Plant growth is stunted when soil water and air conditions in the root zone are inadequate. Similarly, water content in a field beyond its capacity level inhibits growth. Drainage therefore becomes the most viable option for remedying the waterlogged condition and providing an unsaturated root zone for the plant. Drained soils are more productive, amenable to mechanization, especially tillage operations, and ensure a healthier work environment for farm labour.

Soil hydraulic conductivity is the dominant factor in designing subdrainage installation (Ohize, 1999). It influences the quantity of water that can be drained from a unit area of land per unit time. Various approaches, methods and procedures alongside some theories and hypotheses, have been used to determine the value of soil hydraulic conductivity both on the field and in the laboratory. Various physical and mathematical models have also been developed for estimation of soil hydraulic conductivity (Kozeny, 1927 reviewed by Carman, 1937; Fair and Hatch, 1933; Childs and Collis-George, 1950; Marshall, 1957). Kirkham (1946) proposed a method for measuring soil hydraulic conductivity in the field when a water table is present, a condition which adequately reflects the horizontal component of soil hydraulic conductivity, and that Luthin (1973) regards as dominant and most important in drainage engineering.

The soils of Kubani River Valley are fertile. The land is put to production of rice and sugarcane at the lower reaches and arable crops on the high grounds in the rainy season. Tomato, onion and carrot are among the vegetable crops grown in the dry season. Over the years productivity appears to have declined in the area. This has been attributed largely to waterlogging, which can be corrected by subsurface drainage. The present research therefore determines the saturated hydraulic conductivity of soils at site that can be used for the design of a subdrainage works. The two commonly used methods the single auger hole method developed by Hooghoudt and described in Luthin (1973) and the pipe cavity method of Kirkham (1946), were employed.

MATERIAL AND METHODS

The experiment was conducted at a farm along Kubani River Valley, downstream of Ahmadu Bello University dam at Samaru, Nigeria (11°11′ N, 7°38′ E, 686 m above sea level). The vegetation is typically Sudan savanna with scattered shrubs. The weather comprises a hot rainy season in May to October, followed by a prolonged cool dry season in the months of November through March and a short hot dry season in April through June. The soils are predominantly sandy loam to a depth of about 1 m.

The experiment was of factorial design with two methods of determining saturated hydraulic conductivities *in situ* combined with five depths of auger hole below the water table. The depth of hole below the water table and other dimensions are given in Table 38.1. The ten factor-level combinations were randomly laid out in an up-river-bank axis with the first hole dug closest to the water. Four determinations were carried out at each hole. Depth of the impermeable layer was assumed to be infinite. Holes for SAHM and PCM tests were spaced 0.50 m apart to prevent interference of one hole by another and, anyway, the two tests were conducted separately.

Table 38.1 Depth of hole below the water or depth to end of pipe at the five hole locations

Hole no.	*Depth of hole (SAHM)*	*Depth to end of pipe (PCM)*
1	64	47
2	70	54
3	67	48
4	86	65
5	61	44

SAHM:

Radius of auger hole = 38 mm

PCM:

Radius of pipe = 10.7 mm
Radius of cavity = 12.7 mm
Length of cavity = 100 mm

Single Auger Hole Method (SAHM)

The single auger hole method entails boring/augering a hole below the water table, pumping out the water and measuring its rate of rise in the hole from a predetermined depth. The essential equipment used in our test included a 76 mm diameter orchard-type auger, which is best suited for augering in stable soils. A hole scratcher sized to fit the hole was sourced to remove any possible sealing of the side of the hole that resulted from augering. The scratcher was made from hard wood turned on a lathe machine to 50 mm diameter, 150 mm length with nail heads driven into the surface at 1.5 nails per cm^2, carefully ensuring even insertion of the nails into the wood. This part was attached to a 2 m long, 12 mm galvanized iron pipe handle by a welded plate at one end, using three 13 mm wood screws. A 300 mm deep bailer was also constructed from a 0.75 m galvanized pipe. A plastic ping-pong ball on a rubber sitting served as a foot-valve at one end. The rubber sitting rested on a bevelled surface to allow the ping-pong ball to sit snugly on the hole, aided by a column of water to provide a watertight seal as the bailer was withdrawn from the hole. At the other end, a wire string was attached through two 6 mm holes on opposite sides of the bailer for easy withdrawal or lowering into the hole. A stop-watch with digital readout was used to record time intervals between water rises as actuated by the float. A flashlight was used for inspection of the hole when the need arose. A vertical recording board and flat apparatus were constructed from white soft wood, mild steel rod and plastic ball to record the rise of water in the hole. This board, 100 mm × 1000 mm × 30 mm, was cut peg-like at one end to facilitate driving it into the ground next to the hole. Additional stability was provided to the board by a crossboard 300 mm × 1000 mm × 30 mm nailed to it on the edge that rested flat on ground level. A graduated scale of 0-500 mm was taped to the board with masking tape. Two eye-screws were fixed in the middle of the board in an up-down axis 350 mm apart to allow for easy movement of the 6 mm diameter steel rod carrying a pointer. The steel rod had its lower end glued into the 65 mm plastic ball that served as the float. The rod carrying the pointer was free to move up and down a span of 500 mm, thus accommodating the highest possible rise of water in the hole.

The auger hole was bored vertically down to the desired depth, noting any variation in soil texture, colour and structure. Less than full auger capacity of soil was withdrawn from the hole at each boring. On completion, the sides of the holes were well scratched with the scratcher to remove any sealing due to augering. Thereafter, time was allowed for water to move into the hole and rise to static water level. When equilibrium was established, measurements were taken of depth to static water table from the ground surface, total depth of hole and depth from static water table to the bottom of the hole. At the start of data gathering, the recording board and float apparatus were brought close enough to the hole. The hole was bailed out a second time, carefully ensuring that only about half the depth of water in the hole was removed. The apparatus was subsequently lowered into the hole and positioned ready for recording, ensuring that the gauge and pointer were moving freely. As the gauge began to rise from the action of the float, the stop-watch was activated when the pointer reached the initial bottom mark on the recording tape of the board. Using equal rise increments, the time lapse between each was recorded from the stop-watch. The rises over time elapsed were used in the Ernst auger hole formula (Maarsland and Haskew, 1957) to estimate K_{sat} values included in Table 38.2.

Pipe Cavity Method (PCM)

The pipe cavity method essentially required that a tube or pipe be driven into the soil below a water table either with or without a cavity at the end of the tube. The materials/equipment needed for this research

Table 38.2 Saturated hydraulic conductivities K_{sat} (m.day^{-1}) of soils at Kubani and variance analysis of means as affected by depth of hole or end of pipe below the water table at hole locations and determination method

Hole no.	*Method*	*Replications of Reading*					*Variance analysis*
		1	*2*	*3*	*4*	*Mean*	*Location:*
1	SAHM	0.2036	0.1424	0.2036	0.1423	0.1730	
	PCM	0.1685	0.1632	0.1646	0.1504	0.1617	
	Mean	0.1861	0.1528	0.1841	0.1464		0.1674b
2	SAHM	0.3494	0.3631	0.5387	0.3621	0.4033	
	PCM	0.3262	0.3101	0.3220	0.2955	0.3135	
	Mean	0.3378	0.3366	0.4306	0.3288		0.3584a
3	SAHM	0.3482	0.3849	0.3853	0.3800	0.3747	
	PCM	0.3393	0.3477	0.3294	0.3147	0.3328	
	Mean	0.3438	0.3663	0.3574	0.3474		0.3537a
4	SAHM	0.4770	0.3544	0.2980	0.3270	0.3641	
	PCM	0.3513	0.3429	0.3199	0.3323	0.3366	
	Mean	0.4142	0.3487	0.3090	0.3297		0.3504a
5	SAHM	0.4212	0.3802	0.4091	0.4039	0.4036	
	PCM	0.3719	0.3654	0.3833	0.3741	0.3737	
	Mean	0.3966	0.3728	0.3962	0.3890		0.3887a
							Method:
							SAHM 0.3438a
							PCM 0.3037b
							Interaction:
							Location × method NS

Means followed by the same letter within the same treatment group are statistically not significant (NS).

included a 25 mm diameter, 1 m long plastic pressure pipe, a 1.1 m long soil probe, a 12 volt dc electric water indicator which operated as an electrolytic device, a small quantity of table salt (NaCl), a small diameter rubber hose, a soil thermometer and a stop-watch.

The pipe cavity holes, located at least 200 mm near the single auger holes, were carefully bored to a uniform depth of 1 m using the soil probe. As soils were withdrawn from the hole, they were observed for changes in texture, colour and structure in order to validate the observations made during boring of the SAHM test holes. Boring was discontinued when it was confirmed that a cavity of 100 mm would result at the end of the pipe. The pipes were driven in carefully to position. Water level in the pipe was allowed to recover to the level of static water table, then the water table depth was measured. A thermometer attached to a string was lowered into the hole to measure the soil water temperature. Using the flexible rubber hose, water was sucked out of the pipe by a hand-pump. About 5 table salt was introduced into the pipe to increase salt concentration of the water in the hole to enhance detection of water level. The electric water level recorder was lowered in the pipe and rate of water rise from a predetermined depth recorded using a stop-watch. The predetermined depths were varied among holes from 335 to 550 cm. The electric water level recorder was a simple device composed of a flashlight bulb connected to two strands of wire forming the electrodes. The circuit was completed and light flashed when water level in the pipe came into contact with the wire. A graduated tape wrapped around the length of the wire provided a means of measuring the height that water had risen in the tube.

RESULTS AND DISCUSSION

At hole no. 1, located 30 m from the river, the mean saturated hydraulic conductivities of the two methods was 0.1674 m day^{-1}; it was significantly lower than at other locations farther away from the river. For hole nos. 2 through 5, there was no significant difference among the mean K_{sat}. The highest mean K_{sat} of 0.3887 m day^{-1} was obtained at no. 5, located 150 m from the river, which is numerically twice the magnitude at the river-bank. This result is validated by the fact that soil bulk density (given in Table 38.3) of the samples obtained was highest at hole no. 1 with a value of 1.842 g cm^{-3} compared to locations 2 through 5 that ranged from 1.645 to 1.691 g cm^{-3}, an increase of 9–12%. As the riparian area remains seasonally inundated by floods, the soils become compacted, resulting in low saturated hydraulic conductivity. The results contained in Table 38.2 show a significant difference between values of mean K_{sat} obtained by the two methods. The SAHM was significantly higher by 13% with a value of mean K_{sat} = 0.3438 m day^{-1} than the PCM

Table 38.3 Particle size analysis and textural class of the soils at Kubani River Valley, Samaru-Zaria

Hole no.	*Size proportion (%)*			*Textural class**	*Bulk density*
	Clay	*Silt*	*Sand*		*(g cm^{-3})*
1	0	13.74	86.58	Loamy sand	1.842
2	0	7.22	92.78	Sand	1.691
3	0	11.00	89.00	Sand	1.645
4	0	22.60	77.39	Loamy sand	1.686
5	0	21.70	78.3	Loamy sand	1.674

*Based on USDA textural classification chart (redrawn from US Bur. Plant In., Soils and Agr. Eng., 1951).

value of K_{sat} = 0.3037 m day^{-1}. The highest variation of values between the two methods was obtained at hole no. 2 and the least at hole no. 1. The interaction of depth of auger hole with method of determination was not significant (P = 0.05), however.

CONCLUSIONS

From the present work, K_{sat} values determined by the Pipe Cavity Method, though higher than by the Single Auger method, were less affected by depth of hole below the water table indicating some stability with methodology. Similar determinations with the SAHM method varied by as much as 12%. The ease of setting up PCM in the field and the relatively low requirement for materials makes the technique attractive. Inspite of the fact that the internal wall of the auger hole was scratched, sealing of the surface could not be assumed as completely eliminated. This is a drawback for the method.

Acknowledgements

The author thanks the Director, Institute for Agricultural Research, Samaru for supporting this research work.

References

Carman, P.C. 1937. Fluid flow through granular beds. Trans. Inst. Chem. Eng., London 15: 150-166.

Childs, E.C. and Collis-George, 1950. The permeability of porous materials. Proc. Roy. Soc. (London), A201: 392-405.

Fair, G.M. and Hatch, L.P. 1933. Fundamental factors governing the streamline flow of water through sand. J. Amer. Water Works Assoc., 25: 1551-1665.

Kirkham, D. 1946. Proposed method for field measurement of permeability of soil below the water table. Soil Sci. Soc. Amer. Proc., pp. 58-68.

Kozeny, J. 1927. Ueber capillare Leitung des Wassers in Boden. Sitzungsber. Wien. Akad. Wissensch., 136(2a): 271-306.

Luthin, J.N. 1973. Drainage Engineering. Robert E. Krieger Publ. Co., Huntington, NY 250 pp.

Marshall, T.J. 1957. Permeability and the size distribution of pores. Nature. 180: 664-665.

Maarsland, M. and Haskew, H.C. 1957. The auger hole method of measuring the hydraulic conductivity of soil and its application to the tile drainage problem. 3rd Cong. Intern. Comm. Irrig. Drainage. R5, Questions 8: 8.69-8.14.

Ohize, A.O. 1999. *In-situ* determination of hydraulic conductivity of soils at the Kubani River Valley using the single auger hole and pipe cavity methods. A project submitted in partial fulfillment of the requirements for the award of B. Eng. (Agricultural) degree to the Department of Agricultural Engineering, Ahmadu Bello University, Zaria, 41pp.

[illegible]

CONCLUSIONS

[illegible]

ACKNOWLEDGMENTS

[illegible]

REFERENCES

[illegible]

39

EFFECTS OF PESTICIDE TREATMENTS ON YIELD, QUALITY AND QUANTITY OF WINTER WHEAT VARIETIES AND TRACING RESIDUES OF THEIR ACTIVE AGENTS

Zoltan Hegedûs, Marton Jolankai and Zsolt Szentpetery
Szent Istvan University, Institute of Crop Production,
H-2100 Gödöllö, Páter Károly u.l., Hungary
e-mail: Hegedus@fau.gau.hu

INTRODUCTION

Agricultural production has long been a major enterprise in Hungary because more than 60% of the country's land is arable. For centuries, winter wheat was the most commonly cultivated crop. But the first document pertaining to the quality of the wheat produced in various parts of the country was a map drawn up by Johann Treyer in 1885, under commission from the commodity exchange of Wien (Láng, 1997).

Today the quality of Hungarian winter wheat must be assessed in terms of its competitive capability not only for domestic markets, but exportability as well. Aside from good milling and baking quality, the whole process of production should be reviewed to ensure a high-quality, healthy end product. Up-to-date quality assurance systems (e.g. HACCP, ISO) have established criteria for many products, including agricultural (Sósné, 1996). These systems strive to minimalize risks during cultivation through transportation to their destination point to handling and storage and ultimately consumption.

From the technological point of view, for wheat agrochemical interventions are the most critical and hence their impact should be soundly studied. One of the basic principles in sustainable agriculture is the production of goods both safe and healthy for the consumer (van Mansvelt and Mulder, 1993). Hence not only do the effects of chemicals on quality and quantity of wheat production need to be studied, but the residues of their active ingredients accurately traced in the yield (Jolánkai et al., 1996; Mesterházy, 1993).

Milling and baking quality of grains are basically determined by environmental and agrotechnical factors (Birkás, 1996; Kárpáti et al., 1996). Environmental characteristics of a site such as climate, soil, are mainly free of human activity; however, part of them can be modified artificially. The effects of various agronomic methods on milling and baking quality differ (Bedö and Láng, 1997; Vida et al., 1996). It has been said (Pollhamerné, 1981) that quality is genetically determined and that agronomic applications can manifest or worsen, but never improve it. Growing quality wheat therefore requires optimization of growing factors by quality parameters (Ivány et al., 1994). The relationship between pesticide usage and quality of wheat has not been well studied; our experiments will hopefully provide some illumination.

MATERIAL AND METHODS

Field trials were undertaken for five years (1995-2000) in Nagygombos (50 km from Budapest) and one year (1996) in Gödöllö (30 km east of Budapest). The soil type of the experimental field was chernozem with forest residue (Chromic Luvisol) in Nagygombos and mouldy sand in Gödöllö. Experiments were conducted in split plot design with four replications each year. Plot size was 10 m^2 and plots were planted and harvested by plot machines. Average yield and milling quality of winter wheat varieties Alföld 90, Mv 21 and Fatima 2 were studied each year. The quantity of herbicide, fungicide and insecticide residue in the yield were measured throughout the ripening of all varieties. Samples were taken 1, 3, and 6 weeks after spraying.

Fungicide and insecticide treatments were applied according to the principles of Integrated Pest Management. Plant-protection sprayings were done according to gradation of pests (such as *Lema*, *Eurygaster* and *Anisoplia* species) and the appearance of diseases (such as *Puccinia*, *Septoria*, *Erysiphe* species) based on local forecasts.

Herbicides were sprayed after heading (9-10 weeks) much later than usual, to allow infestation, and given in the dosage recommended.

Milling quality was evaluated according to Hungarian Standards (MSZ ISO 531: 1993; MSZ ISO 6645: 1993) by the laboratory of Concordia Ltd., Budapest and the Grain Laboratory of the Institute of Crop Production, Gödöllö. Pesticide residues were detected with an HPLC device.

Yield Average, Quality Assay and Lab Tests

A. Yield of each plot measured and average yield per hectare computed.

B. Milling and baking quality were determined by these parameters:
   — Protein content
   — Test weight
   — Wet gluten content
   — Farinographic value
   — Falling number (Hagberg)

C. Residues for all the pesticides used were determined three times during vegetation (1, 3, 6 weeks after treatments).

Herbicide Treatments

This experiment was carried out to study the effects of different herbicides on both winter wheat and its weed flora. The herbicides used and their active agents were Starane (fluroxipir), Pardner (bromoxynil) and Banvel 480 (dicamba). Dosages are shown in Table 39.1.

Table 39.1 Herbicide treatments and doses

A	weedy	Control
B	hoed	Control
C	Starane	0.8 L ha^{-1}
D	Pardner	2 L ha^{-1}
E	Banvel 480	0.7 L ha^{-1}

Fungicide and Insecticide Treatments

The fungicides used and their active agents were: Bayleton 125EC (triadimefon) and Folicur Top (tebukonazol + triadimefon).

Insecticide and active agent: Bancol 50 WP (benszultap).

The doses of the aforesaid are given in Table 39.2.

Table 39.2 Fungicide and insecticide teatments and doses

Control:	A	
Fungicide:	B	Folicur Top (1 l/ha)
	B	Bayleton (1 l/ha)
Insecticide:	C	Bancol 50 WP (0.75 kg/ha)
Fungicide Insecticide	+ D	Bancol + Bayleton, D Bancol + Folicur

RESULTS AND DISCUSSION

Effects of Fungicide and Insecticide Treatments on Average Yield

Only the Bancol treatment caused a significant loss in yield in 1999 of variety Mv 21 and variety Alföld 90. No significant difference was found in the other treatments.

Effects of Herbicide Treatments on Average Yield

The various herbicides used for plant protection resulted in a lower yield in every wheat variety (Fig. 39.1). With later treatment, however, the differences were not significant. This means that herbicide treatment applied later than usual is an efficient means of weed control in a heavily weeded area.

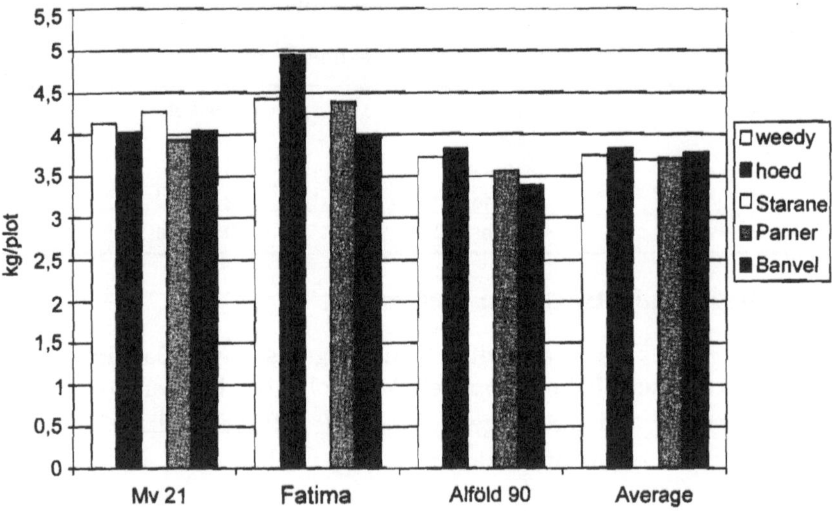

Fig. 39.1 Effect of herbicide treatments on yield of wheat cultivars.

Effects of Fungicide and Insecticide Treatments on Quality Parameters of Wheat Varieties

Changes in protein content, gluten content and farinographic values (Brabender farinograph) were evaluated consequent to treatments. Earlier studies have shown that fungicides and insecticides used appropriately to protect wheat crops during the vegetative period induced only subtle changes in milling and baking quality parameters. Our data from field experiments support these findings. We found no significant difference in test weights among the experimental varieties.

Nor were there significant alterations in wet gluten and protein contents (Table 39.3).

The same is true of farinographic value and falling number.

Table 39.3 Wet gluten and protein content in yield after fungicide and insecticide treatments

Treatments	*Alföld 90*		*MV-21*	
	Protein	Wet gluten	Protein	Wet Gluten
Control	13.46	31.12	13.17	30.6
Insecticide	13.4	31.04	13.3	29.9
Fungicide	13.54	31.24	12.9	28.77
Both	13.28	30.34	13.5	29.9
LSD 5%	0.89	2.33	0.84	3.95

Effects of Herbicides on Quality Parameters of Wheat Varieties

There were no significant changes in test weights. Table 39.4 shows that protein content and wet gluten content were modified by herbicides.

Unlike former hormone-based herbicides, which could markedly enhance the protein content of kernels, chemicals can occasionally modify it. The largest alteration in protein and wet gluten contents was found in Mv 21. Both parameters were higher than control values but nonetheless, no case proved statistically significant. Differences in farinographic unit among the tested varieties are shown in Figure 39.2.

Table 39.4 Changes in wet gluten and protein contents after herbicide treatments

Treatments	*MV-21*		*Alföld 90*		*Fatima-2*	
	Wet gluten	Protein	Wet gluten	Protein	Wet gluten	Protein
Weedy control	31.8	14.17	30.6	13.34	30.7	13.9
Hoed control	32.67	14.7	30.5	13.42	31.4	14.2

(Table Contd.)

Table 39.4 Contd.

Starane	34.1	14.8	30.9	13.48	30	13.6
Pardner	32.9	14.7	30.06	13.28	31	14.2
Banvel	32.7	14	31	13.06	30.2	13.8
LSD 5%	2.87	0.93	3.57	0.87	5.2	1.5

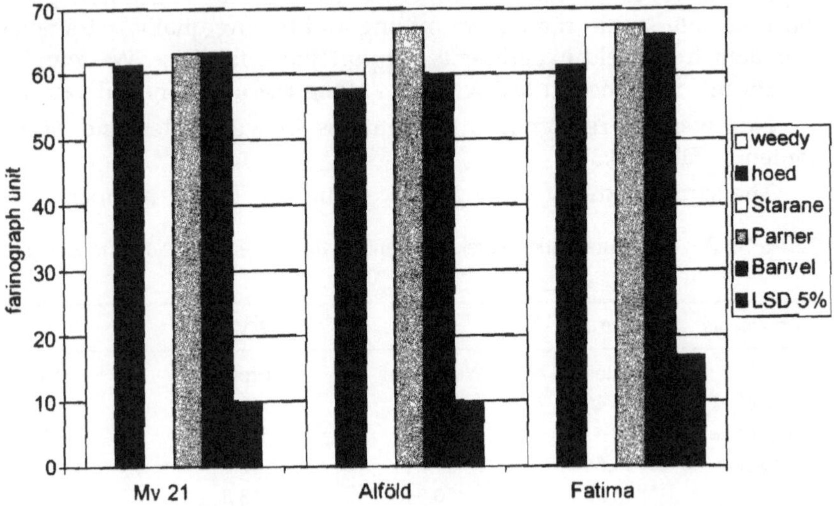

Fig. 39.2 Alteration of farinographic unit due to pesticide treatments.

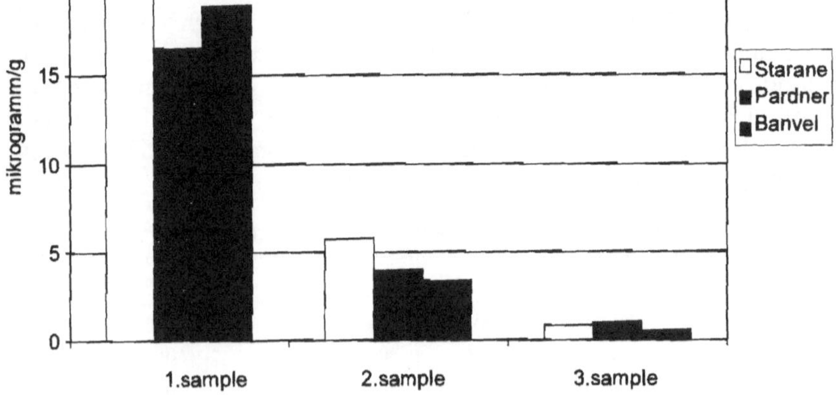

Fig. 39.3 Decrease in herbicide residue content in variety Mv-21.

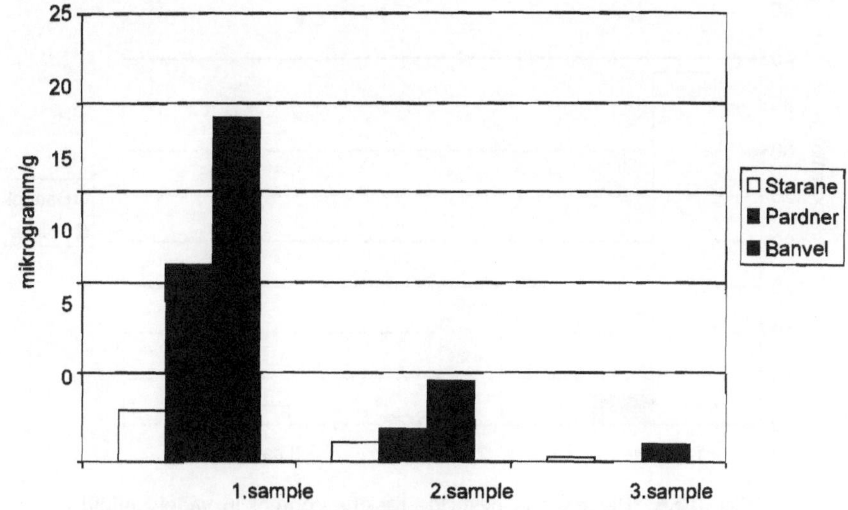

Fig. 39.4 Decrease in herbicide residue content in variety Fatima 2.

Trends in Herbicide Residues in Yield

Figures 39.3 and 39.4 show the herbicide residue contents for variety Mv 21 and Fatima 2 respectively over all years. Residues tended to decrease rapidly during the vegetative period and the level of contamination was very low at harvest.

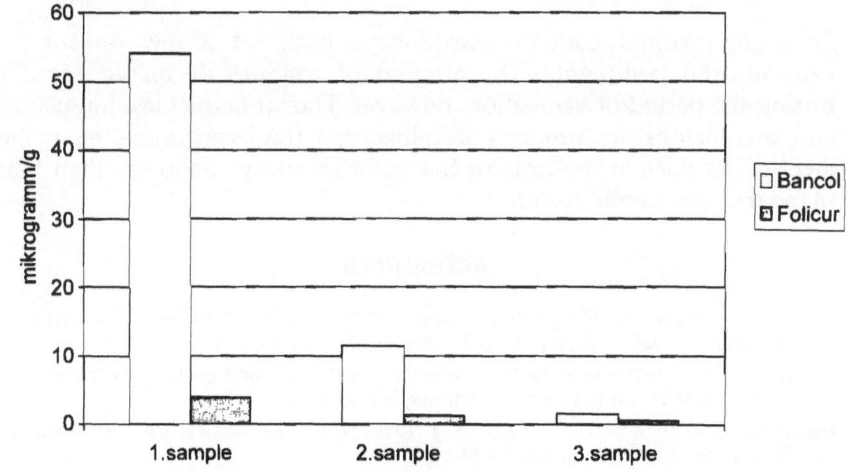

Fig. 39.5 Decrease in pesticide residue content in variety Mv 21.

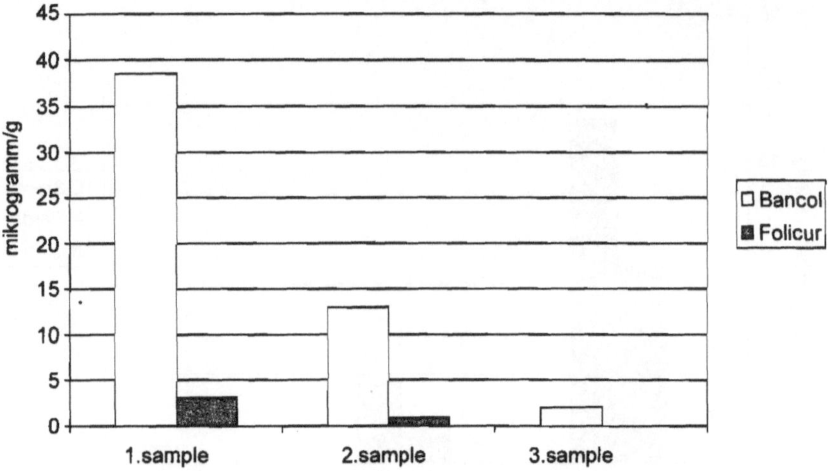

Fig. 39.6 Decrease in pesticide residue content in variety Alföld.

Trends in Fungicide and Insecticide Residues

Figures 39.5 and 39.6 show the residual amounts of fungicides and insecticides 1, 3 and 6 weeks after treatment of variety Mv 21 and Alföld 90 respectively. A decreasing trend similar to that observed for herbicides was noted here also.

CONCLUSION

Pesticide residual content was always high for a few days after experimental treatments. The amount of residues decreased rapidly during the period of vegetation, however. Thus if herbicides, fungicides and insecticides are properly administered (no overdosage, etc.), the level of residues at the time of harvest will always be lower than that mandated for public health.

References

Bedö, Z. and Láng, L. 1997. A minöségbúza termesztése és nemesitése [Growing and breeding of quality wheat]. Agro-21 Füzetek, 97 (14): 8-28. (in Hungarian)

Birkás, M. 1996. Földmüvelés és földhasználat [Tillage and land use]. Egyetemi jegyzet. GATE, Gödöllö. (in Hungarian, with English summary)

Ivány, K., Kismányoky, T. and Ragasits, I. 1994. Növénytermesztés [Crop Production]. Mezögazda Kiadó, Budapest. (in Hungarian)

Jolánkai, M., Szentpétery, Zs. and Szalai, T. 1996. Variety specific weed tolerance—a key to non-chemical weed control. Proc. ESA 4th Cong., Veldhoven-Wageningen, Vol. II, pp. 556-557.

Kárpáti, M., Szentpétery, Zs., Jolánkai, M., Varga, J., Örsi, F. and Simonné, S. L. 1996. Az öszi búza tartalékfehérjéinek és aminosav-összetételének tanulmányozása a szemkialakulás során [Study of pooled protein and amino acids of wheat during filling]. Növénytermelés 45 (3): 255-263. (in Hungarian with English summary)

Láng, I. 1997. A minöség dimenziöi az agrárgazdaságban [Dimensions of quality in the agricultural sector]. Agro-21 Füzetek 97 (14): 3-7. (in Hungarian)

Mesterházy, A. 1993. Resistance to diseases as a factor reducing costs in wheat production, pp. 104-105. In: J. Ángyán et al. (eds.). GATE, Gödöllö.

Pollhamerné, E. 1981. A búza és a liszt minösége [Quality of wheat and flour]. Mezögazdasági Kiadó, Budapest. (in Hungarian)

Sósné dr Gazdag, M. 1996. Minöségbiztositás az élelmiszeriparban [Quality assurance in food production]. Mezögazda Kiadó, Budapest. (in Hungarian)

van Mansvelt, J.D. and Mulder, J.A. 1993. European features for sustainable development: a contribution to the dialogue. Landscape and Urban Planning 27: 67-69.

Vida Gy, Bedö, Z. and Jolánkai, M. 1996. Agronómiai kezeléskombinációk öszi búzafajták sütöipari minöségére gyakoroit hatásának elemzése fökomponens-analizissel [Impacts of various agronomic methods on baking quality of wheat evaluated with factor analysis]. Növénytermelés 45 (6): 453-462. (in Hungarian with English summary)

40

INFLUENCE OF VARYING LIGHT INTENSITIES ON NODULATION OF SOYBEAN (*GLYCINE MAX* (L.) MERRILL)

F.O. Odeleye[1], T.O. Tayo[2], A.O. Togun[1] and R. Dris
[1]University of Ibadan, Department of Crop Protection and Environmental Biology, Ibadan Nigeria.
[2]University of Agriculture, Department of Plant Science and Crop Production, Abeokute, Nigeria.
e-mail: aotogun@skannet.com; *e-mail:* gbengaodeleye@hotmail.com

INTRODUCTION

Environmental factors are particularly important in the cultivation of legumes. Water deficits, temperature, soil acidity and light intensity—all have profound effects on the growth, development and yield of soybean in particular (Hunt et al., 1987; 1990; Kasperbauer et al., 1984). The effect of spectral composition of light on soybean has shown that self-regulation of nodulation in soybean seedlings was affected by FR and R light treatments (Hunt et al., 1987) and that nodulation of soybean seedlings can be altered by variation in the spectral composition of light reflected from different coloured surfaces (Hunt et al., 1985).

However, the intensity of light (the photon flux per unit surface area of leaves) also affects the number of nodules in the roots of soybean, aside from the effect of reduced light intensity in enhancing vegetative growth of the crop. Several workers have studied the effect of light intensity on the growth, development and yield of soybean (Hang et al.,

1984; Odeleye et al., 2001) but very little has been done in the area of effect of reduced light intensity on nodulation of soybean, especially in south-west Nigeria where the crop is usually shaded by taller intercrops during cultivation. The main objective of the present study was to assess the effect of light intensity on nodulation in soybean.

MATERIAL AND METHODS

Three soybean varieties were selected for the study based on maturity period: TGX 1485-1D (early maturing) TGX 845-313D (medium maturing) and Malayan (late maturing). The seeds of these varieties of soybean were obtained from the International Institute of Tropical Agriculture (IITA), Ibadan and sown in pots in 1992. The pots were arranged in split–split plot manner on the roof top of the department of Crop Protection and Environmental Biology, University of Ibadan a field experiment (see below) was also undertaken.

Cage Construction

Cages made of wood, 5 cm × 5 cm, were constructed and covered with 1 or 2 layers of synthetic 1 mm green mesh net obtained from the local market. They were covered on all sides with one or two layers to reduce light intensity by 25% or 50% respectively. Light intensity within and outside the screens was measured with a light meter, Model 4555 Type C (Megatron, England). The values of the light measurements taken on bright and cloudy days in the roof top garden and in the field are presented in Table 40.1.

Cultural Practices

Seeds of the three soybean varieties were sown in plastic pots (5-litre capacity with 20 cm rim diameter), each containing 3.95 kg soil obtained from the plot where the field trial was eventually carried out. Three days prior to this, NPK (15: 15: 15) fertilizer was applied at the rate of 50 kg ha^{-1}, so that 1.47 g fertilizer was added to each pot. Five seeds were sown per pot at a depth of 3 cm. The resultant seedlings were thinned to one per pot, two weeks after sowing.

The pots were arranged in a split-split plot layout on the roof top garden of the department of Crop Protection and Environmental Biology, University of Ibadan. The soybean varieties served as the main plots, stage of growth as subplots and the light regimes the sub-subplot, with five replications. Randomization was done according to the procedures

Table 40.1 Mean squares of significant 2-way interactions for nodules of soybean plants grown under reduced light intensities at three growth stages.

Pot experiment

		Nodule Number			*Nodule Dry Weight*		
Source of variation	df	2 WAT	4 WAT	6 WAT	2 WAT	4 WAT	6 WAT
Variety × stage	4	1960.91*	1462.96*	534.02*	0.41*	0.18*	0.13 ns
Variety × light	4	394.95*	591.56*	36.24*	0.08*	0.21*	0.09 ns
Stage × light	4	1612.96*	1592.58*	678.18*	0.47*	0.80*	0.49*

Field experiment

		Nodule Number			*Nodule Dry Weight*		
Source of variation	df	2 WAT	4 WAT	6 WAT	2 WAT	4 WAT	6 WAT
Variety × stage	4	1433.56*	1115.06*	19119.49*	11.88*	12.55*	22.00*
Variety × light	4	178.86*	19.29*	88.39*	0.15*	0.36*	0.82*
Stage × light	4	167.51*	225.29*	14.48*	0.59*	0.17*	0.10*

*Significant at the 5% level of probability; ns-not significant. WAT-weeks after treatment.

of Gomez and Gomez (1984). Two light regimes, 75 and 50% daylight (L1 and L2 respectively), were imposed (using the constructed cages) for two weeks on the soybean varieties at the vegetative stage (V4-5, when the plants had four to five nodes on the main stem beginning with the unifoliolate node) early flowering stage (R2, when the plants flower at the node immediately below the uppermost node with completely unrolled leaf) and at the pod-filling stage (R5, when the seeds are beginning to develop and can be felt when squeezed at one of the four uppermost nodes with completely unrolled leaf) (Fehr and Caviness, 1977). Plants receiving 100% light (L0) intensity served as control.

Sixty five pots of the early-maturing variety and eighty pots each of medium-and-late maturing varieties were removed at the vegetative stage of growth and arranged inside the cages. At the flowering and pod-filling stages of growth, 50 pots of early maturing and 65 pots each of the medium-and-late maturing varieties were transferred into the cages. The plants remained inside the cages for two weeks.

Sampling and Measurement

The treated plants were removed from the cages after two weeks and returned to 100% light intensity. Sampling at fortnightly intervals commenced a day after expiration of treatment imposition. Five plants per treatment were evaluated for leaf number, node number, nodule number, root dry weight, nodule dry weight and total dry weight. The sampled plants were carefully removed so that nearly all the roots and all the nodules were recovered. Dry matter of the different plant parts (roots, nodules and total plant dry weight) was determined after oven drying the samples at 80°C for 48 h.

Field Experiment

This was a split-split plot. Each main plot measured 4.2 × 6 m, each subplot 1.4 × 6 m and each sub-subplot 1.4 × 2 m. Treatments were randomly allocated to main plots, subplot and sub-subplot as done in the pot experiment.

Planting and cultural practices were as described by Odeleye et al. (2001). The light reducing cages were transferred to the field and imposed on the soybean plants as they reached V4-5, R2 and R5 stages of growth. The cages were left for two weeks in each treatment. Sampling commenced one day after cage removal from each treatment and fortnightly thereafter. Data collection was done as in the pot experiment. The plants were carefully uprooted with a hand-trowel so that all the roots and nodules were recovered. The nodules were counted and their dry weight taken.

RESULTS

Varietal Effects

The early-maturing variety (V1) produced the lowest number of leaves. The numbers of nodes of the medium-(V2) and late-(V3) maturing varieties were similar but significantly higher than that of V1(see Fig. 40.1). The highest number of nodules was produced by V3. V1 also produced more nodules than V2.

Effect of Stage of Growth

All the cultivars initially developed significantly more leaves after treatment at the pod filling stage (S3) than other growth stages. Later, plants treated at the vegetative stage (S1) produced more leaves than plants treated at the other two stages. The number of nodes of S2 plants was similar to that of S3 plants. The lowest number of nodules was recorded at the S3 stage because nodule degeneration was most rapid in plants treated at S3. Shortly after treatment, plants treated at R5 (S3 plants) had significantly more nodules than plants treated at vegetative and early flowering stages (S1 and S2 plants). Two weeks after treatment

(WAT), the order was S2 > S1 > S3. At 4 WAT and beyond, the order was S1>S2>S3.

Effect of light intensity

Imposition of the various light treatments caused no significant differences in number of leaves. The number of nodes in L1 and L2 plants were similar but significantly more than in L0 plants for most of the growth period after treatment. The highest number of nodules was produced by L2 plants, which had significantly more nodules than L0 and L1 plants.

Interactions

The variety × stage of growth interaction was highly significant for number of nodules throughout the post-treatment period. The variety × light intensity interaction was significant for leaf number and nodule number up to 2 WAT. The stage of growth × light intensity interaction was highly significant for all vegetative parameters measured, including nodule number up to 4 WAT.

Dry Matter Accumulation

Varietal effect

The root dry weight of V2 was significantly higher than those of V3 and V1.Nodule dry weight of V1 was higher than those of V2 and V3 at 6 WAT.

Effect of stage of growth

The root and nodule dry weights were initially lower in plants treated at S1 than those treated at S2 and S3 but later the dry weights of these parameters increased in S1 plants such that they had the highest values of these parameters at penultimate and final harvests (Fig. 40.1).

Effect of light intensity

The nodule dry weights of plants under L1 was the highest at 4 and 6 WAT but was significantly the lowest at 2 WAT. The root dry weight of L0 plants was the lowest throughout the post-treatment period while L2 plants had the highest at final harvest.

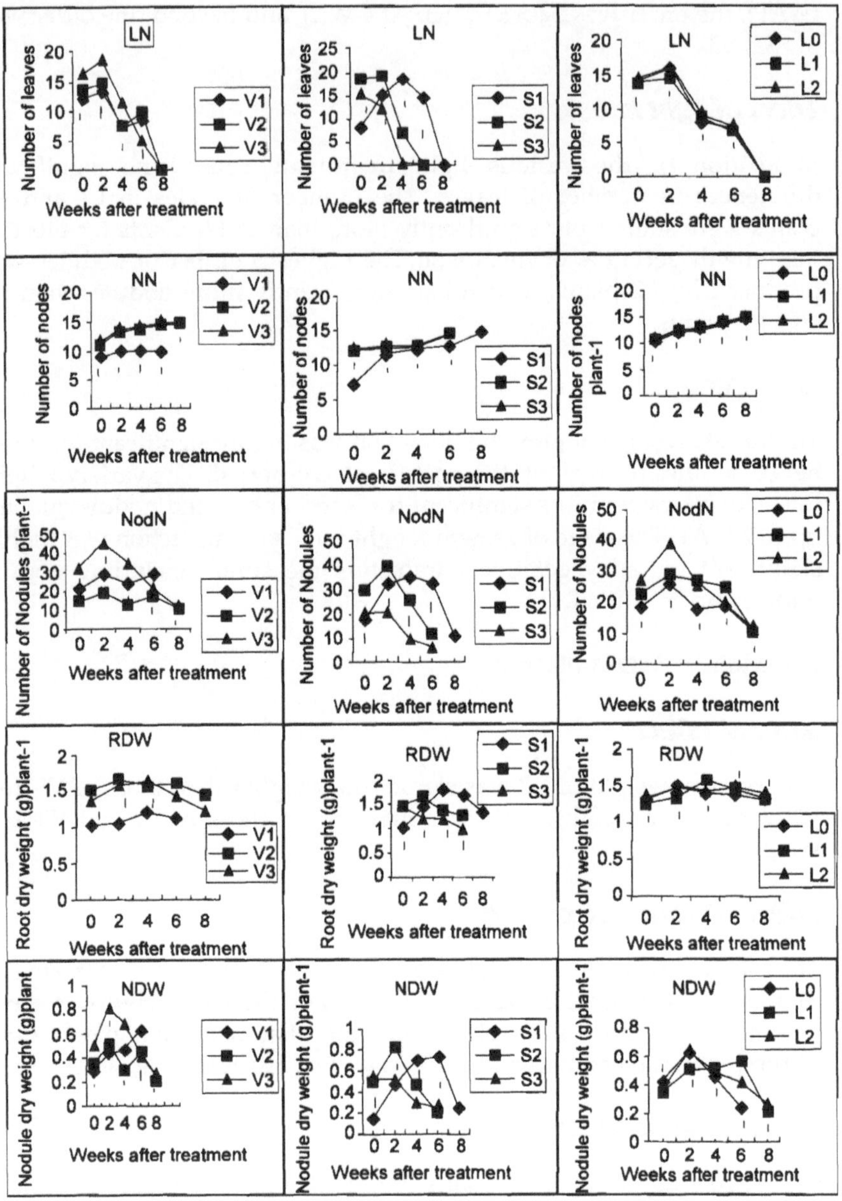

Fig. 40.1 Effects of variety, growth stage and light intensity on growth and nodulation of soybean.

Interactions

The variety × stage of growth interaction was highly significant for root and nodule dry weight until 4 WAT. At 6 WAT, the interaction was not significant for root and nodule dry weights. The variety × light intensity interaction was significant for root and nodule dry weights at 2 WAT. The stage of growth × light intensity interaction was significant for root and nodule dry weight throughout the period of growth after treatment.

Field experiment

As in the pot experiment, the late-maturing variety had significantly more nodes and nodules than V1 and V2 for most of the period after treatment.

Plants treated at S2 and S3 had more nodules initially than S1 plants but consequent to rapid degeneration of nodules in plants of S2 and S3, S1 plants eventually had the highest number of nodules at final harvest.

At final harvest, L0 plants had more nodules than L1 and L2 plants. The L1 and L2 plants, however, initially produced more nodules than L0 plants. The L1 and L2 plants had a higher number of nodes than L0 plants until final harvest.

Interactions

The variety × stage of growth interaction was highly significant for number of nodules and other vegetative parameters considered shortly after treatment and at 4 WAT. It was significant for leaf number and nodule number at 6 WAT.

The variety × light intensity interaction was significant for leaf number and number of nodules at 4 and 6 WAT.

The stage of growth × light intensity interaction was highly significant for number of leaves and number of nodules one day, two weeks and four weeks after treatment.

Dry matter accumulation

The stem root and nodule dry weights of V2 were significantly higher than those of V1 and V3 throughout the post-treatment period.

S2 plants had the highest values of root and nodule dry weight at 4 and 6 WAT.

Nodule production was initially highest in L2 plants but at final harvest these plants had the lowest nodule dry weight.

Interactions

The two-way interactions of variety × stage of growth, variety × light intensity as well as stage of growth × light intensity were significant for nodule dry weight one day, two weeks and four weeks after treatment.

RESULTS AND DISCUSSION

The highest number of nodules were produced by plants grown under reduced light intensities shortly after treatment, both in pots and in the field. There was also a general enhancement of vegetative growth under reduced light intensities. The need for increased fixation of nitrogen under reduced light intensities may be responsible for this initial increase in number of nodules since under reduced light intensity there is usually a reduction in nitrogen in shaded leaves (Salisbury and Ross, 1992).

Nodules in plants grown under lower light intensities appear to degenerate rapidly, as shown by the results of this study. The implication of this is that nodules are often reduced in number during pod development when they are probably needed most. This may have contributed to the eventual yield reduction of the soybean varieties as reported by Odeleye and colleagues (2001). On the contrary, plants grown throughout in the open ultimately had the highest number of nodules, particularly at final harvest. This may imply fixation of nitrogen by the maturing plants, which may have enhanced plant yield.

Since the lowest number of nodules and a corresponding lowest dry weight of nodules were obtained at the pod-filling stage of growth, it follows that light reduction actually caused the most rapid loss of nodules at this stage. It is therefore likely that at the pod-filling stage of growth, competition for assimilates by the developing pods and sustenance of metabolism by the vegetative parts (including the nodules) was fiercer and that the amount of assimilates reaching the nodules was not enough for sustaining their metabolism, hence the most rapid degeneration occurred at this stage of growth.

The highest number of nodules produced by the late maturing variety was probably because of the greater vegetative growth/vigour of this variety compared to the medium- and early-maturing varieties and not because of higher efficiency of carbon assimilation, since the late-maturing variety actually had the lowest Net Assimilation Rate (data not shown). The early-maturing variety produced the highest number and dry weight of nodules at maturity (6 WAT) compared with the medium- and late-maturing varieties. This implies that the early-maturing variety used in this study was capable of retaining nodules until maturity so that its reproductive growth could be continually enhanced due to probable continuous fixation of nitrogen when most useful to the plant.

In the field, however, the late-maturing variety had the highest number of nodules from 4 WAT until 8 WAT. Despite this, the medium maturing variety had a higher nodule dry weight than to the late-maturing variety. The difference in nodulation of the varieties in the field and pots can be ascribed to the ability of the medium- and late-maturing varieties to fully express their potential in the field in terms of nodule production and vegetative growth (unrestricted root growth). However the inability to retain nodules until maturity persisted in the field showing that this is an innate attribute of these two varieties. The significant variety × light intensity interaction for nodule number and dry weight suggests that the varieties responded differently to the various light regimes in terms of nodule production. In conclusion, the two levels of light reduction (50 and 75% daylight) adversely affected nodulation in soybean and the most devastating effect on nodulation was obtained when light reduction was imposed at R5 (pod-filling stage).

References

Board, J.E. and Harville, B.G. 1993. Soyabean yield component responses to a light interception gradient during the reproductive period. Crop Sci. 33: 772-777.

Fehr, W.R. and Caviness, C.E. 1977. Stages of soybean development. Iowa Agric. Exp. Stn. Spec. Report 80. Ames, the IA.

Gomez, K.A. and Gomez, A.A. (2nd ed.) 1984. Statistical Procedures for Agricultural Research. John Wiley and Sons, pp. 12-14.

Hang, A.N. McCloud, D.E. Boote, K.J. and Duncan, W.G. 1984. Shade effects on growth partitioning and yield components of peanuts. Crop Sci. 24: 109-115.

Hunt, P.G., Kasperbauer, M.J. and Matheny, T.A. 1987. Nodule development in a split root system in response to red and far-red light treatment of soybean shoots. Crop Sci. 27: 973-976.

Hunt, P.G., Kasperbauer, M.J., and Matheny, T.A. 1990. Influence of *Bradirhizobium japonicum* strain and Far-Red/ Red canopy light ratios on nodulation of soybean. Crop Sci. 30: 1396-1308.

Hunt, P.G., Sojka, R.E., Matheny, T.A. and Wollum, A.G.II. 1985. Soybean response to *Rhizobium japonicum* strain, row orientation and irrigation. Agron. J. 77: 720-725.

Kasperbauer M. J., Hunt, P.G. and Sojka, R.E. 1984. Photosynthate partitioning and nodule formation in soybean plants that received red or far red-light at the end of the photosynthetic period. Physiol. Plant. 61: 549-554.

Odeleye, F.O., Togun, A.O. and Tayo, T.O. 2001. The effect of light intensity on the growth development and yield of soybean (*Glycine max* (L.) Merrill) in southwest Nigeria. African Crop Sci. J. (in press).

Salisbury, F.B. and Ross, C.W. 1992. *Plant Physiology*. Wadsworth Publ. Co. Belmont, CA, pp. 439-462.

41

PRELIMINARY RESULTS OF FLOROVIT AND PINIVIT APPLICATION FOR CONTROLLING FRUIT SET

A. Basak[1], E. Szwonek[1] and B. Dyki[2]

[1]Research Institute of Pomology and Floriculture, ul. Pomologiczna 18, 96-100 Skierniewice.

[2]Research Institute of Vegetable Crops, ul. Konstytucji 3 Maja 1/3, 96-100 Skierniewice.

e-mail: abasak@insad.pl

INTRODUCTION

A peculiar feature of fruit trees is their tendency towards irregular fruiting. It can be observed that under certain conditions an excessive number of flowers set fruit. As a result, due to a deficiency of nutrients necessary for their development, fruits do not achieve the desired size. Consequently, there is a need to reduce the number of fruits per tree by thinning. Such a procedure makes harvesting of high-quality fruit every year possible. New methods for controlling fruit set are continually sought to make thinning more reliable, and the results of the procedure repeatable and safe for the environment.

Most often the procedure involves chemical treatment with synthetic chemicals restricting fruit set or stimulating drop of weaker fruitlets. However, none of the chemicals used to date has been ideal for that purpose (Wertheim, 1998). The present study looks at the specific, above all film-forming properties of the liquid fertilizers Florovit Uniwersalny and Pinivit commonly used to control fruit set in apple trees. It was assumed that spraying apple-trees with these chemicals during flowering, after fertilization of central flowers (which produce top quality fruits) would make it impossible for the flowers developing later (which generally produce fruits of lower quality) to set fruit. The authors could find no information in the available literature on a similar method of controlling fruit set. This procedure was therefore employed in an attempt to combine thinning with a simultaneous supply of additional nutrients to trees.

MATERIAL AND METHODS

The study was carried out in the Experimental Orchard of the Research Institute of Pomology and Floriculture in Dabrowice near Skierniewice in the year 2000. The study objects were 4-year-old apple trees cv. Jonagold grafted on M.26 rootstock. The trees were growing in five neighbouring rows at a spacing of 4 m between rows and 2.5 m between trees in a row. For the purpose of controlling fruit set the following chemicals were used: Florovit Uniwersalny (composition in %: N 11.4, P 4.0, K 13.9, Mg 2.3, Ca 6.4, B 0.025, Cu 0.05, Mo 0.027) and Pinivit (N 5.6%, P 0.6%, K 2.7%, Fe 600 mg L^{-1}, Mn 3000 mg L^{-1} and Zn 150 ml L^{-1}).

Five treatment methods were evaluated. For the first spray the recommended concentrations of the fertilizer solutions were used. Provocative concentrations were used for the second spray. The two types of concentrations were 3 and 10% v/v respectively, for both Florovit and Pinivit. Treatment of the trees sprayed with the fertilizers only once was carried out on 26 April, while the trees sprayed twice were first treated on 26 April and then 2 May. The study involved the following experimental combinations:

1) Control—trees sprayed with water;
2) Florovit—one spray (3% solution);
3) Florovit—two sprays (3% and 10% solutions);
4) Pinivit—one spray (3% solution);
5) Pinivit—two sprays (3% and 10% solutions).

Each combination included 24 trees growing in separate rows. The trees were sprayed with Florovit and Pinivit once, after fertilization of central flowers (about 3 days after they had opened up), or twice. In the latter case, some of the trees were sprayed with the fertilizers again approximately a week later. Eight trees of similar size and flowering intensity were selected from each experimental combination for detailed assessment.

The following observations and measurements were made.

- Degree of fruit set was established. For that purpose all flower clusters per tree were counted at the beginning of flowering. Then the number of fruits per tree was determined twice: prior to June drop (1.06) and after the drop period (30.06). On the basis of the ratio of number of fruits to number of flower clusters the initial fruit set (before June drop) and the final fruit set (after June drop) were determined.
- On the third day after spraying, which was probably two days after fertilization of lateral flowers, flowers were taken for microscopic assessment of the progress of their pollination and fertilization.
- Following their removal, the pistils were macerated in IN NaOH for 10 seconds, then rinsed in distilled water and stained in 0.06% aniline blue for 30 seconds. The material was analysed using a fluorescent light microscope with OG_1 and BG_3 filters.
- Number of pollen grains on the stigma and number of pollen tubes in the style were determined (Dyki, 1978). The examined specimens were photographed.
- Towards the end of June (27.06) indicator leaves were taken for analysis to determine N (total), P, K, Mg and Ca content. The total nitrogen content was determined using Kjeldahl's method, while ICP was used to determine the other elements.
- During harvest the yield was determined as number and weight of apples collected from each tree separately.
- Size of apples on individual trees was established by calculating the mean weight of 1 apple (in grams)
- Internal quality of apples at harvest was determined on the basis of their firmness (pounds cm^{-2}) and refraction (%); 20 apples were taken from each tree for this evaluation.

Results relating to fruit set, yield, apple size, firmness and refraction were subjected to statistical analysis using Duncan's t-test. The significance of the differences between the experimental combinations was determined at the significance level $a = 0.05$.

RESULTS AND DISCUSSION

A very large number of pollen grains (too many to be counted) was found on the stigmas of flowers of the control (unsprayed) trees and the large number of pollen tubes growing into their ovaries indicated fertilization (Table 41.1).

On the other hand, pollination and fertilization of flowers on trees sprayed with the fertilizers were significantly worse. Differences in the effectiveness of the fertilizers tested were found. Pinivit had a greater effect on these processes than Florovit. Applying the chemicals twice caused greater changes than spraying only once.

In the case of a single treatment with Florovit, a very large number of pollen grains was found on the stigmas of the pistils. This indicated extensive pollination of flowers. However, significantly fewer pollen tubes were growing into the ovary of these flowers compared to control. In some flowers, pollen tubes were present in the style but had not grown into the ovary. In flowers sprayed with Florovit twice, pollination and fertilization were in general worse than after the single treatment, and the stigmas were less receptive.

Treatment with Pinivit, irrespective of the number of sprays, accelerated the aging process in flowers. It clearly caused poorer pollination and fertilization than the application of Florovit. Spraying with Pinivit twice caused desiccation of the stigmas. This indicates that the product may have phytotoxic properties.

Florovit and Pinivit treatments caused an increase in total nitrogen and potassium content in the indicator leaves. It was also established that leaves of the trees sprayed with these fertilizers contained slightly less magnesium and calcium than those of control trees. Phosphorus concentration in leaves was not modified by Florovit and Pinivit within the statistically proven range (Table 41.2).

The fertilizers did not have a significant effect on fruit set, either at the initial and final or before and after June drop. However, trees sprayed with Pinivit twice produced fewer fruits than those treated in other ways and the untreated control (Table 41.3).

Spraying the trees with Florovit and Pinivit had no significant effect on the size of apples as measured by their average weight. These fertilizers are used for foliar fertilization of trees.

The absence of any effect of the treatments on fruit size means that the apples on all the trees in this experiment had adequate conditions for development and additional fertilization was not needed.

Due to late spring frosts, fruiting by all the trees in the experiment was poor. The trees had a large number of leaves per fruit, which ensured supply of sufficient quantities of nutrients to the apples.

Trees sprayed with Pinivit twice had the lowest yield in terms of both number of apples per tree at harvest and fruit weight in kg per tree. The differences, however, were not proven statistically.

Table 41.1 The effect of foliar fertilizers Florovit and Pinivit on pollination and fertilization of flowers in apple trees cv. Jonagold

Treatment method	*Successive flowers*											*Remarks*
	1	*2*	*3*	*4*	*5*	*6*	*7*	*8*	*9*	*10*	*11*	
Control (untreated)	–	X+	+	X+	X+	X+	X+	X+	X+	–	X+	Generally, abundant pollen in pollination aboundant pollen tubes in fertilization and
Florovit One spray	–	X+	X+	X+	+–	+–	X+	X+	X+	X+	X+	Extensive pollination, but fewer pollen tubes than in control
Florovit Two spray	+–	+–	X+	+–	X+	X+	X+	X+	+–	X+	X+	Generally, reduced pollination and fertilization, stigmas less receptive.
Pinivit One spray	X+	+–	–	–	X+	+–	–	–	U	U	U	Poor pollination and fertilization, aged flowers
Pinivit Two sprays	–	+–	X+	X+	–	–	X+	U	U	U	U	Poor pollination and fertilization, aged flowers; desiccated stigmas.

Explanations:

X : Uncountable pollen grains on the stigma.

+– : Pollination present, but pollen tubes do not enter the ovary

+ : Pollination present and pollen tubes in the ovary

– : Absence of pollination due to desiccated stigma or because fertilization had already taken place.

U : Wilted flowers.

Table 41.2 Macroelement content in the indicator leaves of apple trees treated with foliar fertilizers Florovit and Pinivit (samples taken on 27 June)

Treatment method	*N (total) content % (p.d.w.).*	*P*	*K*	*Mg*	*Ca*
Control (untreated)	1.92 a	0.18	1.12 a	0.31 b	1.90 d
Florovit One spray	2.11 c	0.18	1.58 c	0.25 a	1.02 a
Pinivit Two sprays	1.98 ab	0.18 1 .s.d.	1.48 b	0.27 a	1.35 b
Florovit Two sprays	1.99 ab	0.16	1.22 a	0.29 ab	1.52 c
Pinivit One spray	2.06 bc	0.17	1.39 b	0.26 a	1.27 b

p.d.w.–per cent dry weight; 1.s.d. – lowest significant difference.

Table 41.3 Effect of foliar fertilizers Florovit and Pinivit on fruit set, yield and quality of apples cv. Jonagold.

Treatment method	*Number of fruits from 100 flowers clusters*		*Yield*		*Mean weight of one apple g*	*Firmess pounds cm^{-2}*	*Ref-raction %*
	Before drop	*After drop*	*kg/tree*	*Fruits/tree*			
Control (untreated)	23.629	20.759	11.89	43.39	272.09	12.5	15.9
Florovit One spray	23.439	16.109	9.39	31.49	277.89	12.6	15.9
Florovit Two spray	26.509	19.609	10.79	40.69	251.69	12.5	15.8
Pinivit One spray	23.929	18.959	12.09	38.69	300.09	12.8	15.6
Pinivit Two sprays	14.479	10.349	6.39	20.19	280.49	10.8 1.s.d.	16.5 1.s.d.

Duncan's t-test: a = 0.05.

Apples from the trees sprayed with Pinivit twice had firmness but concomitantly higher refraction. Thus this treatment method had an effect on the 'internal' quality of apples despite the absence of differences in their average weight.

Until now fruit set has been controlled by destroying weaker flowers by spraying them with chemicals having burn properties, e.g. urea (Basak and Dawydko, 1996), ammonium thiosulphate (Basak, 1998), Endothall (Basak and Krzewińska, 1998) or ethephon—a compound releasing ethylene and accelerating the aging process and flower and fruit drop (Wertheim, 1998).

For thinning purposes, wetting agents containing polymers of fatty acids have also been used in the past, which destroy the surface of the cuticle on the stigma making it unreceptive (Basak and Faron, 2000). The present studies, in particular those pertaining to microscopic assessment of flowers, suggest a new method of controlling fruit set—by covering the stigmas with a film-forming substance which would make pollen germination difficult.

The fertilizers Florovit and Pinivit are commonly used in commercial orchards for foliar fertilization of trees. The results presented in this paper indicate that using these products for foliar fertilization can at the same time inhibit pollen germination and ovary fertilization. This usually leads to a reduction in fruit set. Florovit and Pinivit contain natural compounds. Using them to control fruit set would thus constitute an ecologically safe procedure (Basak, 2000).

References

Basak, A. 1998. Tiosiarczan Amonu (ATS)-nowy preparat do przerzedzania zawiazkòw jabloni w okresie kwitnienia. Mat. XXXVII Ogòlnopölskiej Nauk. Konf. Sad., Skierniewice 25-27.08. 1998, pp. 333-338.

Basak, A. 2000. Przerzedzanie zawiazkòw w warunkach proekologicznej produkcji owocòw. Mat. Konf. "Proekologiczne technologie produkcji owocòw", Skierniewice, kwiecieñ 2000: 64-70.

Basak, A. and Dawydko, T. 1996. Wykorzystanie moczika i zwiazkòw mocznikopodobnych do przerzedzania zawiazkòw jablek. Mat. II Ogòlnopolskiego Symp. "Nowe Ro`sliny I Technologie w Ogrodnictwie", Pozna`n 17-19.09.1996, 1: 108-112.

Basak, A. and Krzewińska, D. 1998. Wyniki badán z zastowaniem preparatu Endothall do przerzedzania zawiazkòw owocowych. Nauk. Akad. Roln. w Krakowie, 57(1): 369-373.

Basak, A. and Faron, A. 2000. Wyniki badán z zastosowaniem preparatu Armo Thin do przerzedzania zawiazkòw owocowych jabloni. Zesz. Nauk. Akad. Roln. w Krakowie, 71: 219-222.

Dyki, B. 1978. Zastosowanie metody fluorescencyjnej do badania samoniezgodnósci linii hodowlanych kalafioròw. Biul. XXI: 267-272.

Wertheim, S.J. 1998. Chemical thinning of deciduous fruit trees. Acta Hort. 463: 445-462.

42

SNAPDRAGON (*ANTIRRHINUM MAJUS*) CUT FLOWERS—PAST AND FUTURE IN HORTICULTURE

U. Moor and S. Vabrit
Estonian Agricultural University, Institute of Horticulture,
Kreutzwaldi 64, 51014, Tartu, Estonia
e-mail: ulvi26@hot.ee

ADVANTAGES FROM NATURE—BIOLOGY AND ORIGIN

Genus *Antirrhinum* belongs to family Scrophulariaceae. For horticultural purposes, *Antirrhinum majus* cultivars have been divided into three groups: tall – 60-90 (120) cm, medium—20-40 cm and dwarf—up to 20 cm (Bailey, 1969; Cavendish, 1969; Fogg, 1972). Since only tall cultivars can produce high-quality cut flowers, only this group of snapdragons is discussed here.

Single florets attach to a long flower stalk, thus ensuring a relatively long vase life of cut flowers. Various sources indicate that snapdragons should be harvested when 1/3, 1/2, 2/3 of the florets have opened, the remaining buds opening during vase life. One research work (Lee et al., 1995) has indicated, however, that optimal harvesting for snapdragon is the stage of 7-9 opened florets.

Antirrhinum majus L. is an indigenous wild plant in the Mediterranean and southern part of Europe, which means it is well adapted to the hot, sometimes dry climate of these regions. It also tolerates relatively poor soil. Already in 1757, the Englishman John Hill had written regarding snapdragon: 'Every gardener would laugh were we to give instructions for cultivating this plant, as it is sowing itself as a weed and can even grow in an old stonewall' (cited by Coats, 1956). Snapdragon can grow in almost every type of soil, except heavy clays. Its need for fertilizers is low (Bevasqua and Mellano, 1993) but it can tolerate rather high amounts of nitrogen without reduction in vase life of cut flowers (Wilcke, 1996). This general modesty in terms of ecological/ nutritive conditions notwithstanding, the plant appears to have some specific requirements. B-deficiency in the soil during the growing period has been reported to affect the life of cut flowers while Ca-rich soils cause desiccation and abscission of the tip of the main branch (Salunkhe et al., 1990). Evidently these defects are cultivar(s)-related as B-deficiency in our soil in 1999 was not reflected in the quality of our cultivars.

Antirrhinum majus is a sun-loving plant. Even though artificial light is generally not supplied in commercial snapdragon production, its beneficial effect on the quality of cut flowers has been reported by some researchers (Armitage, 1992; Salunkhe et al., 1990).

The colour range of snapdragon cut flowers is wide, extending to every colour except blue. Much research has been done on the pigments of snapdragon flowers and it was discovered that purple, caused by cyanidin, predominates over other colours in snapdragons (Misiha and El-Atawy, 1992; Toki and Kawanishi, 1992).

Problems with *Antirrhinum majus* in the Past: Historical Résumé

A major problem in cultivating snapdragons has been rust caused by the fungus *Puccinia antirrhini*. The disease first manifested in America, from where it reached the United Kingdom in 1933 (Cavendish, 1969) and spread rapidly. Scientists tried for rust-resistant cultivars but new clones of the disease developed simultaneously. Years of research by the Royal Horticultural Society in Wisley, England finally concluded at the end of the 1970s with 16 rust-resistant cultivars. Most snapdragon cultivars nowadays are rust resistant. *Antirrhinum majus* has always been susceptible to mostly fungal diseases. Thus irrigation should always be straight to the soil so that water cannot reach the leaves and flowers. With dip-irrigation used in greenhouses these days, watering should not be a problem.

Another very important characteristic of cut flower quality is vase life. The vase life of various taxa is one among other factors dependent on plant response to ethylene. Since *A. majus* is very sensitive to ethylene, it was at a great disadvantage compared to ethylene-insensitive cut flowers such as *Dendranthema* in the past. But ethylene ceased to be a problem when impressive effects of silver ion in inhibiting its action were first reported by Beyer (1976). Soon dramatic improvements in vase life were being reported for many cut flowers, including *Antirrhinum majus* (Reid and Wu, 1992; Borochov and Woodson, 1989). But as time passed, it became clear that use of silver thiosulphate (STS) was not the perfect solution to the problem because it contains silver, an environmental pollutant. With further progress in science, 1-MCP (1-methyl cyclopropene) was synthesized (Sisler, 1991; Sisler and Serek, 1997). It provided as much protection as STS, preventing ethylene-induced bud and flower abscission, leaf abscission and flower senescence while concominantly being an environment-friendly product. Patented, 1-MCP is commercially available at present under the name EthylBloc.

Gravitropism is described as one of the post-harvest problems of snapdragon cut flowers. Stem-bending away from gravity, an important cause of quality loss in spike-type flowers such as *Gladiolus*, *Kniphofia* and *Antirrhinum majus*, is the effect of redistribution of auxin in response to gravity (Woltering et al., 1991).

Negative geotropism in such flowers can be prevented by a pulse treatment with naphthyl pthalamic acid (NPA), an inhibitor of auxin transport. This treatment has proven very effective with snapdragons (Serek and Reid, 2000). In summer 2001, we used certain chemicals to overcome gravitropic response in snapdragon cut flowers. To our amazement, flower stalks stored horizontally for two hours in a dark room at a temperature of +8°C showed no curvature (unpubl. data).

SEVEN YEARS RESEARCH ON *ANTIRRHINUM MAJUS* IN ESTONIAN AGRICULTURAL UNIVERSITY

In 1995, the main purpose of experiments was to ascertain the suitability of some snapdragon cultivars offered by Benary Seed Company for open-field growing. In 1996-1997, the same cultivars were grown in the experimental garden summer flower collection and the influence of weather on flowering observed. From 1998 to 2001, attention was focused mainly on the influence of differences in cultivars, weather conditions, post-harvest storage temperature, some commercial preservatives and post-harvest water stress on the post-harvest life of snapdragon cut flowers.

- Time from sowing to flowering depends on the cultivar and its tolerance to weather conditions. From 1995-1998, this time was less influenced by weather in cv. Gefüllte Scarlet Supreme. Weather conditions exerted the greatest influence on flowering of Riesen-Vorbote cultivars. Gefüllte cultivars flowered earlier, 94-99 days after sowing, with Riesen-Vorbote lagging behind at 97-108 days.

- Gefüllte cultivars were the most productive and had the greatest amount (74%) of first-quality cut flowers.
- Post-harvest life of snapdragon cut flowers was influenced first and foremost by the amount of precipitation, average temperature and sum of radiation received in the ten-days period prior to harvest (Moor et al., 2000).
- The preservative solution 'Flower Fresh' prolonged the post-harvest life of all snapdragon cut flowers by five days. The most significant positive effect was seen with Gefüllte cultivars, however, which stayed fresh 8-10 days longer.
- Snapdragon cut flowers preserved in a refrigerator at +4.5°C stayed fresh two weeks longer than those maintained in a mean room temperature of +14°C.
- Gefüllte cultivars Scarlet Supreme and Superjet proved the most suitable for cut flower production in open fields in Estonia. These cultivars had early flowering, high-quality cut flowers and were sufficiently weather tolerant. High-quality cut flowers with a long post-harvest life could also be produced with Riesen-Vorbote cultivars Purpursamt, Weiss and Leuchtscharlach, but the risk of quality loss always remained when weather conditions were not optimal.
- *Antirrhinum majus* cv. Riesen-Vorbote Purpursamt and Rosa cut flowers tolerate post-harvest water stress up to 24 h without significant difference in vase life afterwards. Water stresss for more than 24 h leads to a dramatic increase in amount of dropped florets (R^2 = 0.9835) and dry storage more than 48 h results in irreversible dehydration (unpubl. data).

ANTIRRHINUM MAJUS AS A CUT FLOWER FOR FUTURE CONSUMER DEMANDS

Changes in consumer tastes have been noticed since the beginning of the present century. Consumers are not only interested in good-looking flowers with a long vase life, but also in their mode of production. In the long term, consumer tastes will be strongly influenced by their increasing concern about quality of the product and environmental protection, resulting in a greater desire for disease- and pest-resistant plants that need no spraying for instance (Cadic, 2001).

Since 1990, snapdragon has been the subject of numerous molecular biology and gene technology studies. It was one of the first ornamental crops to be genetically transformed for many purposes (Godwin et al., 1991; Hoshino et al., 1998; Robinson and Firoozabady, 1993). Genetic control of flower development by homeotic genes was based on studies of *Antirrhinum majus*, leading to beneficial knowledge for further research in many other horticultural crops (Schwarz-Sommer, 1990).

Many of the advantages of *A. majus* cut flower come from the aforesaid characteristics: initially a good natural base and subsequently much human endeavour. It is a strong plant with modest demands for cultivation conditions. It can readily be grown in open land, even as far north as Estonia (58°41′ NL, 25°55′ EL). Thus it can be grown in open land at low cost in large areas in many countries, not year-round, but still as a complement in the flower market in summer to imported flowers. It will not be long before we have to give thought to our unrestorable natural resources, when home country plants will be valued even more.

Many local species do not have the advantage of research granted to snapdragons and hence will be discarded due to cultivation problems. It has happened before. Improving these species is one way of obtaining new ornamentals in horticulture (Hentig, 1995).

Given its extensive research, it can be said that nearly all the problems of *Antirrhinum majus* have been resolved and this flower can be grown as an environment-friendly product. Sometimes high quality and environment compatibility are difficult to conjoin. Furthermore, definitions of quality differ; for example some are visible such as length, ripeness and size while others are not visible but can be very decisive in satisfying a consumer, such as sensitivity to leaf and bud drop, vase life and ornamental value (Noordegraaf, 1995).

Antirrhinum majus seems to possess characteristics that can meet both visible and invisible quality demands. For instance, very few cut flowers can tolerate transportation without water without experiencing difficulty in rehydration subsequently. Hence wetting agents are used adding to cost. If not hampered by some other disadvantage, *Antirrhinum majus* is readily rehydrated even after 36 hours of water stress. It has been mentioned that flowers such as wax flowers, introduced to cultivation from an arid environment, have greater immunity to the adverse effect of transient water stress (Mayak et al., 2001). We might suggest that such could be the case for *Antirrhinum majus* also.

One of the most significant factors in the survival of a given genotype is its ability to produce seeds and propagate itself. For most flowers, including *Antirrhinum*, this depends on pollination. Once an individual plant has developed such a feature that confers an advantage in pollination, the gene encoding it will establish rapidly. It has been suggested that it is possible for a character to be fixed in a population by pollinator preference in a single generation (Galen, 1996). As the research of Glover and Martin (1998) has shown, *Antirrhinum majus* genotypes, which contain the *MIXTA* gene, have achieved this advantage. These genotypes produce petals with concical-papillate epidermal cells (Kay et al., 1981). Such cells of the inner epidermis of petals act to increase the pollinator attention received by the plant (Glover and Martin, 1998).

In conclusion, we may say that *Antirrhinum majus* has and is developing by itself to survive and attract more attention from both the beneficial organisms in nature and from consumers in the ornamental market.

References

Armitage, A.M. 1992. Ornamental Bedding Plants. CAB International, Wallingford, England, 154 pp.

Bailey, M. 1969. A Dictionary of Cut Flowers. 367 pp.

Bevasqua, R.F. and Mellano, V.I. 1993. Sewage sludge compost's cumulative effects on crop growth and soil properties. Compost Sci. Utilization, pp. 34-37.

Beyer, E.M. Jr. 1976. Silver ion: a potent antiethylene agent in cucumber and tomato. HortSci. 11: 195-196.

Borochov, A. and Woodson, W.R. 1989. Physiology and biochemistry of flower petal senescence. Hortic. Rev. 11:15-43.

Cadic, A. 2001. Breeding goals for new ornamentals. Acta Hortic. 552: 75-83.

Cavendish, M. 1969. Encyclopedia of Gardening. GB. pp. 57-58.

Coats, A.M. 1956. Flowers and Their Histories. GB. pp. 24-25.

Fogg, W. 1972. Dictionary of Annual Plants. GB. pp. 70-71.

Galen, C. 1996. Rates of floral evolution: adaptation to bumblebee pollination in an alpine wildflower, *Polemonium viscosum*. Evolution 50: 120-125.

Glover, B.J. and Martin, C. 1998. The role of petal cell shape and pigmentation in pollination success in *Antirrhinum majus*. Heredity 80: 778-784.

Godwin, I., Todd, G., Ford-Lloyd, B. and Newbury, H.I. 1991. The effects of acetosyringone and pH on *Agrobacterium*-mediated transformation vary according to plant species. Plant Cell Repts., pp. 671-675

Hentig, W.U. 1995. The development of "new ornamental plants" in Europe. Acta Hortic. 397: 9-29.

Hoshino, Y., Türkan, I. and Mii, M. 1998. Transgenic bialaphos-resistant snapdragon (*Antirrhinum majus* L.) produced by *Agrobacterium* rhizogenes transformation. Sci. Hortic. 76(1-2): 37-57.

Kay, Q.O.N., Daoud, H.S. and Stirton, C.H. 1981. Pigment distribution, light reflection and cell structure in petals. Bot. J. Linn. Soc., 83: 57-84.

Lee, J.S., Kim, I.A. and Sin, I.M. 1995. Effects of harvesting stage, preservative, and storage method on vase life and flower quality of snapdragon. J. Korean Soc. Hortic. Sci. 36(6): 926-942.

Mayak, S., Meir, S. and Ben-Shade, H. 2001. The effect of transient water stress on sugar metabolism and development of cut flowers. Acta Hortic. 543: 191-197.

Misiha, A. and El-Atawy, Y.S. 1992. The genetic system controlling flower colour in *Antirrhinum majus* L. as identified by diallel analyses. Bull. Fac. Agric. Univ. Cairo, pp. 333-352.

Moor, U., Vabrit, S., Pae, A. and Dris, R. 2000. Influence of preharvest conditions on postharvest life of *Antirrhinum majus* cut flowers. In: Proc. of IIF-IIR Conf. Improving Postharvest Technologies of Fruits, Vegetables and Ornamentals. Vol. II. pp. 776–780

Noordegraaf, C.V. 1995. How to obtain and maintain quality. Acta Hortic. 405: 123-131.

Reid, M.S. and Wu, M. J. 1992. Ethylene and flower senescence. Plant Growth Regul. 11: 37-43.

Robinson, K.E.P. and Firoozabady, E. 1993. Transformation of floriculture crops. Sci. Hortic. 55(1-2): 83-99.

Salunkhe, D.K., Bhat, N.R. and Desai, B.B. 1990. Postharvest Biotechnology of Flowers and Ornamental Plants, NY, pp. 22-126.

Sansavini, S. 2000. Funding, management and organization: the key issues facing research in horticulture today and tomorrow. Acta Hortic. 514: 207-215.

Schwarz-Sommer, M. 1990. Genetic control of flower development by homeotic genes in *Antirrhinum majus*. Science (*Washington*), pp. 931-936.

Serek, M. and Reid, M.S. 2000. Use of growth regulators for improving the postharvest quality of ornamentals. In: A. S. Basra (ed.). Plant Growth Regulators in Agriculture and Horticulture. Food Products Press, Binghampton, NY.

Sisler, E.C. 1991. Ethylene-binding components in plants, pp. 81-99. In: A.K. Mattoo and J.C. Suttle (eds.). The Plant Hormone Ethylene. CRC Press, Boca Raton, FL. USA.

Sisler, E.C. and Serek, M. 1997. Inhibitors of ethylene responses in plants at the receptor level: Recent developments. Physiol. Plant. 100: 577-584.

Toki, K. and Kawanishi, H. 1992. Survey of peonidin glycosides in garden snapdragons and changes in anthocyanin contents during floral development. J. Jap. Soc. Hortic. Sci., pp. 989-995.

Wilcke, C. 1996. Haltbarkeit von Sommerblumen. Deutsche Gartenbau 24/96:1376-1379.

Woltering, E.J., Somhorst, D., Beekhuizen, J.G. and Spekking, W.T.J. 1991. Ethylene biosynthesis, carbohydrate metabolism and phenylalanine ammonia-lyase activity in gravireacting Kniphofi flower stalks. Acta Hortic. 298: 99-109.

43

VEGETABLE CONSUMPTION AND DIETARY LIFESTYLE

Nkiru T. Meludu
University of Ibadan, Department of Agricultural Extension and Rural Development, Nigeria

INTRODUCTION

Vegetables are produced in Nigeria not only for consumption, but also for income generation. Even though research gave little attention to the improvement of vegetable production in the 1970s, Nigerian farmers were able to produce enough to meet domestic requirements. Establishment of the National Horticultural Research Institute (NIHORT), Ibadan, Nigeria marked the beginning of research on vegetables in the country. The Institute initially focused on classification and selection of crops according to national and/or regional importance (Denton and Olufolaji, 2000). Attention then turned to the improvement of indigenous crops and the introduction of some vegetable species from various countries, even India, China and South America (Denton and Olufolaji, 2000).

Root vegetables have always been a priority in Nigeria because regarded as a 'man's' crop (Ojehomon, 2000). The per capita availability of root vegetables in the nation has led to high caloric consumption. Women, however, play a major role in leafy and fruit vegetable production in the country (Meludu et al., 1998; 1999). Whatever the crop, there is need for greater dissemination of information pertaining to proper harvesting procedures for various vegetables, their specific handling techniques, nutritional benefits and diversification in preparation and cooking methods. This article highlights the need for awareness of the nutritional benefits of leafy vegetables, in particular proper and better methods of cooking them to maximalize the nutrients essential for good health.

Vegetable Supply and Nutritional Intake

The food consumption pattern in Nigeria has not really changed much over the years in terms of nutritional content—high intake of carbohydrates and low intake of proteins. The typical Nigerian diet consists of yam or cassava as staples, which are nutritionally inadequate in proteins, fat, carbohydrates, vitamins and mineral content. Nonetheless, carbohydrates exceed the recommended level in the Nigerian diet. Provision of fresh vegetables in daily meals would greatly benefit the health of anyone living in the tropics (Gibbered and Gibbered, 1979). Fresh vegetables (cellulose) induce peristaltic movement and thereby indirectly aid digestion and prevent constipation. Contrarily, old and very hard/coarse vegetables can cause indigestion (O'Reilly-Wright, 1985). Excessive consumption of vegetables can result in soft/frequent stools. Thus moderation must be exercised, while always bearing in mind that absence of vegetables in the diet or a deficiency in certain ones can be dangerous to health.

Vegetables have been classified as follows: root or tuber (cassava, yam, carrot, potato and others); leafy (spinach, *Amaranthus*, *Celosia*, lettuce etc.); fruit vegetables (tomato, cucumber, okra, pepper, melon etc.); bulbous (onion, shallot, cabbage etc.); and pulse or legume (broad bean, soybean, peas, groundnut). Although these vegetables are often available in local markets, they have generally been transported long distances and hence never fresh. Spoilage occurs during transportation due to poor storage facilities, lack of preservation and inadequate processing technology. In fact, vegetables rich in nutrients are not always available locally and hence expensive, especially European types (carrot, cucumber, cabbage and lettuce) are imported. Most vegetables contain protein, carbohydrates, mineral salts, vitamins and water. Leafy vegetables are rich in calcium and iron, with folic acid in the very dark ones. Protein and fat are high in pulses but low or only as trace elements in leafy vegetables.

In planning a healthy diet, the consumer must know the nutritive composition of foods and in particular vegetables (Table 43.1). This will ensure meeting the nutritional needs of each family member.

Table 43.1 Average composition per 100 g edible portion of selected vegetables

Selected Vegetables	*Water (ml)*	*Calories (Kcal)*	*Carbohydrate (g)*	*Fiber (g)*	*Calcium (mg)*	*Iron (mg)*	*Vit. A (I.U)*	*Vit.C (mg)*
Carrot	90	33	7	0.8	40	0.7	3000	6
Cucumber	96	12	2	0.5	15	0.3	0	10
Eggplant (Leaves)	03	22	4	1	10	1	0	6
Spinach (Leaves)	85	48	5	1.5	260	4	3000	10
Pumpkin (Leaves)	91	28	4	0.8	80	2.5	1000	5
Cabbage (Leaves)	93	23	4	0.8	40	0.5	30	100
Okra	90	33	6	1	70	1	150	25
Onion	87	48	11	0.5	30	0.5	0	10
Pepper (Red, green)	90	37	6	1	20	1	100-1200	150
Tomatoes	94	20	4	0.6	5	0.4	250	25

Source: adapted from Okeke and Nnabuchi (1996).

Diversifying Vegetable Consumption

Given the diverse array of vegetables and their nutritive composition, Table 43.1 warrants some study. Some change in the volume of each type of vegetable consumed is necessary for good health. With meat and fish becoming more expensive every year, many Nigerians are changing their eating pattern to include vegetables. This is reflected in the high consumption of soybean in the country. Although some of the nutritive vegetables are expensive, eating habits must change to inclusion of 2-3 servings of leafy and/or fruit vegetables every day. The practice of eating adequate meals daily is crucial to healthy living.

Information on Vegetables and Consumer Purchasing Decisions

The role of parents (women in particular) is important here as children need to be educated about nutritive needs and types of vegetables that meet them. Good eating habits and an adequate diet suitable to age and lifestyle are also important.

Where adequate information on foods and vegetables for consumption is not available, a proper system of education needs to be developed. This could be through different media—TV, magazines for instance. With increasing urbanization, the place of consumption and the place of production are increasingly distanced. Awareness of what is involved in agricultural crop production is important in maintaining price stability in vegetable supplies, making them available at the lowest price possible, while concomitantly protecting individual livelihood.

Achieving a Nutritionally Adequate Diet

To achieve a nutritionally adequate diet, buying techniques and choice of vegetables should be taken into consideration. The consumer should make sure that the vegetable to be bought is fresh, crisp and good in colour. If cost is a major factor, certain vegetables are thereby ruled out. It should be borne in mind that in-season vegetables are cheaper. Hence a planting and harvesting calendar for leafy and fruit vegetables (Table 43.2) should be made available to consumers. Such information could be incorporated in the annual national calendar/almanac.

Table 43.2 Planting and harvesting calendar

Vegetables	*Planting Northern Nigeria*	*Planting Southern Nigeria*	*Harvesting in the North*	*Harvesting in the South*
Carrot	June-September	April-September	December-January	December-January
Cucumber	June-September	March-September	October	October
Eggplant	June-September	March-September	October	October
Spinach	May-September	March-October	January	February
Pumpkin	June-July	May-July	October-Nov.	October-Nov
Cabbage	June-August	May-August	November	November
Okra	June-July	September-October	November	February-March
Onion	June-August	April-May	October-Nov.	September-Oct.
Lettuce	May-September	March-October	Dec.-January	November-Dec.
Peas	July	Unsuitable	Oct.-November	Unsuitable
Tomatoes	May-September	March-September	Nov.-December	December

Source: Adapted from Gibbered and Gibbered (1979).

To conserve the nutrients of cooked vegetables until actually eaten, care must be taken in the preparation of leafy/fruit vegetables. Leafy/fruit vegetables left in room temperature and/or in light, can lose much of their folic acid, vitamin B_2 and about 50% of vitamin C in just a few hours (Davis, 1979). Vegetables should not be soaked because some of the nutrients readily pass into the water and flavour is also lost since they contain aromatic oils that are not true oils. Nutrients are also lost during peeling, cutting and shredding of vegetables. Overcooking of vegetables should be avoided as it destroys some of the nutrients, such as vitamin C, spoils colour and diminishes flavour. In fact, it is better to eat most vegetables raw to realize all the nutrients in the right proportion. Several vegetables in a salad are more nutritional than eating large helpings of just one. To cook vegetables, boiling, stewing and steaming are best.

Securing a Stable Vegetable Supply

In a country where adequate electrical supply is lacking, storage of vegetables is difficult. Frozen vegetables retain their nutritive value better than any other form of preservation. However, frozen vegetables should be cooked while frozen to preclude loss of nutrients during defrosting. Leafy vegetables should be salted after cooking and just before eating because salt affects their iron content. Stability of vegetable supply can be achieved through increased vegetable production. Women should be encouraged to plant home gardens for vegetable production, especially the urban women, and not just ornamentals.

CONCLUSION

Vegetable growing in Nigeria is so rapid that a quick succession of different kinds of leafy vegetables can be raised. A mixture of vegetables served in a tasty soup, sauce/stew can provide a good main course meal. Vegetables combined with a small quantity of beef or poultry or fish can provide a substantially adequate diet—at reasonable cost. Above all, people must be educated to the importance of good diet and establish ideal eating habits based on a variety of vegetables.

References

Davis, A. 1979. Let's Cook it Right. Unwin Paperbacks, London. pp. 307-364.

Denton, O.A. and Olufolaji, A.O. 2000. Nigeria's Most Important Vegetable Crops: In Agronomy in Nigeria. A book on the theory and practice of Agronomy as it has been in the last 33 years. Compiled and designed by MO. Akoroda. pp. 85-93.

Gibbered, A.V. And Gibbered, V.L. 1979. Gardening Notebook for the Tropics. Longman Group Limited London. pp 55-88.

Meludu, N.T. Yasunobu, K. and Ono, T. 1998. Biodiversity and the Role of Women in Forest Conservation. Japanese Journal of Crop Science 67: 434-435.

Meludu, N.T., Ifie, P.A., Akinbile, L.A. and Adekoya, E.A. 1999. The role of Women in Sustainable Food Security in Nigeria: A case of udu Local Government Area of Delta State. Journal of Sustainable Agriculture 15: 87-97.

Ojehomon, O.O. 2000. Agricultural Development in Nigeria 1950-2000. My Experience: In Agronomy in Nigeria. A book on the theory and practice of Agronomy as it has been in the last 33 years. Compiled and designed by MO. Akoroda. pp 12-14

Okeke, SUN. And Nnabuchi, F.C. 1996. Fundamental Principles in Nutrition. Midas Touch Press Limited/ Onitsha. pp. 39-46.

O'Reilly-Wright, Enid 1985. The Student's Cookery Book Oxford University Press, New York. pp. 77-120.

44

INFLUENCE OF EPIGEAL EARTHWORMS ON QUALITY OF COMPOSTS AND THEIR EFFECT ON CROP YIELDS, SOIL FERTILITY AND BIOLOGICAL ACTIVITIES

M.C. Manna*, S. Jha and J.K. Saha
Indian Institute of Soil Science, Nabibagh, Berasia Road,
Bhopal, 462038, India
**e-mail: mcm@iiss.mp.nic.in*
e-mail: samanta@iiss.mp.nic.in

INTRODUCTION

The floral and faunal life of the deciduous forest of Indian semiarid tropics are disturbed by accidental forest fires during summer that also affect the bio-equilibrium, which in turn has an adverse effect on forest vegetation. Recycling deciduous forest litter appears to be viable alternative for forest conservation, thereby minimalizing environmental degradation. In cultivated arable soil, burning crop residues is commonly practised by Indian farmers, which causes a decline in nutrient status and biological activities in soils. It is also known that Indian soils are often low in organic matter, which plays an important role in soil quality and productivity. Availability of organic matter could be enhanced through compost. Hence it is important to develop a suitable technique for preparation of compost in the shortest time possible and to evaluate its quality on crops and soil health.

Some species of epigeal earthworms have the potential for hastening the decomposition process. However, the reproductive cycle of earthworms is influenced by both environmental conditions and the quality of food materials available (Meentmeyer, 1978; Satchell, 1967; Edwards and Heath, 1975; Lee, 1983). Such faunal activities and their combination for waste recycling needs thus to be evaluated in order to better utilize the potential indigenous species and to design a package of practices that are technically feasible, economically viable and readily acceptable to forest managers and farmers. The present study was undertaken to (i) develop a simple and effective system of decomposing wastes in the shortest possible time (vermicomposting) using three species of epigeal earthworms, (ii) study the population dynamics of the three earthworm species under different substrate regimes in a vermicomposting system, and (iii) compare the application of composts with chemical fertilizer on grain yields of crops and soil biological activity in Typic Haplustert.

MATERIAL AND METHODS

The site is characterized by semiarid and subhumid climate, dry summer, hot rainy season, warm autumn and cool winter in Central India (9° 53'S, 83°34' W at an altitude of 662 m from mean sea level). The mean annual rainfall is 1280 mm. Maximum temperature is usually 42°C in summer and the minimum temperature in winter 11°C. The method of vermicomposting, records in faunal activities, compost maturity index, crop yields and soil biological activities are described below.

Vermicompost

Vermicompost was prepared in situ during March-June of 1998 with three forest litters teak (*Tectona grandis)*, mohwa (*Madhuka indica)* and palas (*Butea monosperma)*, and three crop residues wheat straw (*Triticum aestivum*), soybean straw (*Glycine max* (L) Merr.) and mustard straw (*Brassica juncea*). Open permanent pits of 4 m and 3 m × 1 m were made for this purpose. All the pits were made under a shade about 8 feet high from the ground to exclude rainfall and extreme temperature. The bricks wall constructed above the floor were perforated at about 0.10 m dia (6 no. per m^2) for proper aeration. The perforated wall was blocked with nylon screen (100 mesh) in such a way that the earthworms could not emerge from the pits. Decomposed farmyard manure (aged 2 months) was spread on the bottom of the pit to a thickness of about 3-4 cm. This was followed by layers of each waste in the ratio of 1:1 (w/w). A second layer of FYM was then applied, followed by another layer of waste in the same ratio and ultimately a height of 1 m (weighing about 500 kg material) was attained. Three epigeal earthworms, viz. *E. fetida, P. excavatus* and *D. bolaui* were used as a surface epigeic organic matter decomposer since our earlier studies had shown their distribution to be predominant in this region. Moisture was maintained up to 65-75% throughout the decomposition period and jute bags (gunny bags) spread uniformly on top of the materials to maintain of proper moisture and temperature. Watering was done by sprinkler irrigation during the decomposition period. Initially, the materials were allowed to decompose for 21 days to stabilize the initial temperature. They differed in chemical properties, as shown in Table 44.1. Each species was inoculated separately for each forest litter to compare the efficiency of the three epigeal earthworms and in all, 54 pits (3 species × 6 organics × 3 replications) were maintained for this experiment. The three epigeal earthworms were reared separately by the bed method in a screenhouse before inoculation was done. In this method, decomposed bedding material of cattleshed waste was mixed with soil (1:1, w/w on a dry weight basis). At the start of the experiment, the mixture was inoculated with 10 adult worms (1.19 ± 0.2 g each) per kg waste material in each compost pit and a total of 5000 worms used per pit. The materials were allowed to decompose for 16 weeks. Samples 0.5 m^2 area were removed from each pit at 4, 8, 12 and 16 weeks after commencement of the experiment to count the earthworm population of the individual species by the hand-sorting method suggested by Lee (1985). After counting the earthworms, they and the materials were returned to the pit.

Maturity Index of Compost

A sample was taken from each treatment at 4, 8, 12 and 16 weeks after maturation to analyse the maturity index (index of decomposition) parameters. Chemical analyses were done to study the maturity of compost through simple chemical parameters, viz. total organic carbon (TOC) and total nitrogen (TN) by automated CHNS-analyser (Model-Vario EL; CHNS-analyser, Germany), water-soluble carbon using a material: hot water (1:10) extractant: organic carbon by the chromic acid digestion method of Walkley and Black (1934), water-soluble carbohydrates by the anthrone method (Brink et al., 1960) and cation exchange capacity (CEC) by Ba^{2+} exchange (Lax et al., 1986). To determine lignin and cellulose fractions in the compost, the acid–detergent fibre (ADF-sulphuric acid lignin) pre-extraction (Rowland and Robert, 1994) method was used. Available nitrogen (NH_4-N and NO_3-N) was determined by the method of Bremner (1965) using an auto Kjeldahl analyser. Available phosphorus was analysed using the Olsen extractant

Table 44.1 Chemical characteristics of the waste materials used for composts[a]

	Tectona grandis (*Teak*)	*Madhuka indica* (*Mohwa*)	*(Bulea) monosperma* (*Palas*)	*Triticum aestivum* (*Wheat*)	*Glucine max (L.) Merr.* (*Soybean*)	*Brassica juncea L.* (*Mustard*)
Lignin (%)	10.23	11.31	12.21	6.83	11.81	8.94
Cellulose (%)	28.18	30.23	31.16	32.31	37.34	36.97
Lignin/Cellulose ratio	0.36	0.37	0.39	0.21	0.32	0.24
WSC (%)	4.8	3.9	4.0	3.7	5.1	3.4
WSCarbo(%)	6.9	6.1	5.3	7.3	8.3	5.1
TOC(%)	42.13	43.32	48.40	48.93	50.00	46.80
TN(%)	1.25	0.80	1.23	0.52	1.12	0.56
C/N ratio	33.70	41.38	39.34	94.0	44.6	54.6
Total P (%)	0.135	0.096	0.041	0.14	0.19	0.21
Total K (%)	0.70	0.41	0.62	0.42	0.63	0.42

WSC-water-soluble carbon, WSCarbo-water-soluble carbohydrates.

[a]Estimation of chemical parameters based on dry weight of material and values given the mean of three replications.

with 0.5 M sodium bicarbonate at pH 8.5 (1:20 material: extract) (Olsen et al., 1954). Available K was determined by flame photometry (Hanway and Heidal, 1952).

Experiment and Crop Culture

During the wet season soybean cv. J 10 was sown in the first fortnight of June in a narrow spacing of 45 cm at a depth of 5-6 cm with 120 kg ha^{-1} of seed. Vermicompost was applied at the rate of 5 Mg ha^{-1} on soybean. No chemical fertilizer was applied to the soybean crop in the compost treated plots. The effect of composts vis-a-vis 100% recommended doses of chemical fertilizer (N_{25}:$P_{26.2}$:$K_{16.7}$) application was then compared. All treatments were tested in a completely randomized block design with three replications. Soybean was harvested in the first fortnight of October, 1998.

After the soybean harvest wheat cv. WH-147 was sown in the first fortnight of November with a row spacing of 23 cm in each plot. Wheat received 60, 13.1 and 16.7 kg N, P and K ha^{-1} in the plot where only composts were applied to soybean and compared with 100% recommended dose of inorganic to wheat (N_{120} : $P_{26.2}$: $K_{33.3}$). Phosphorus and potassium as single superphosphate and muriate of potash respectively were broadcast to both crops just before seed sowing whereas nitrogen in the form of urea was applied in two splits in soybean (1/2 basal and 1/2 at 30 days after sowing) and three splits in wheat (1/2 basal, 1/2 at crown root initiation and 1/4 at active tillering stage). Other recommended cultural practices were followed to establish good crops. The N and P concentration in soybean and wheat grain at harvest were determined and N and P uptake computed.

Biological Activities

Microbial biomass carbon was determined by the chloroform–fumigation extraction method of Vance and colleagues (1987). A moist soil sample was collected at one-year interval. Twenty grams soil were fumigated with ethanol-free chloroform for 24 h. A non-fumigated soil sample was also run as control. Both samples were extracted with 0.5 M K_2SO_4 and carbon was estimated as described by Vance and colleagues (1987). The SMBC was computed by the following equation: SMBC = 2.64 Ec, where Ec is the extractable carbon from fumigated and non-fumigated soil. The concentration was expressed as mg kg^{-1} soil. Soil respiration was determined by incubating moist soil samples (50 g, 60% field capacity) in an airtight jar with a beaker containing 10 ml 0.5 M

NaOH for 10 days. The evolved CO_2 was estimated by titration of excess NaOH with 0.25 N HCl after addition of $BaCl_2$ to precipitate CO_3^{-2} ions. The concentration of CO_2-carbon was expressed as mg CO_2-carbon/ 100 g/10 d. Dehydrogenase activity was assayed by the method of Casida and coworkers (1964) and the concentration expressed as µg TPF g^{-1}/24h.

Statistical Analysis

Data collected for the various studies were subjected to variance analysis appropriate to the design as given by Snedecor and Cochran (1967). Significance of treatment differences was ascertained with the 'T' test. Significant differences among treatments were compared with differences at the 5 and 1 per cent level of probability.

RESULTS AND DISCUSSION

Characterization of Different Litter Sources

The highest nitrogen concentration was recorded in teak litter and the lowest in wheat residue. Among the three forest litters palas leaf litter contained higher lignin and cellulose than either mohwa and teak leaf litter and thus maintained high lignin/cellulose ratio (Table 44.1). Overall, the total content of mineral nutrients (N, P and K) was higher in forest litters compared to crop residues, except for soybean straw. There was no significant correlation between nitrogen content and lignin per cent in the materials used.

Population Dynamics of three Epigeal Earthworms under Different Food Materials

The different waste materials used during the 16 weeks of decomposition affected the survival and reproduction rate of the three epigeal earthworms (Fig. 44.1). A continuous increase in worms population was observed in all three forest litter composts up to 12 weeks. The highest reproduction rate was recorded in *E. fetida* in all these composts, followed by *P. excavatus* and *D. bolaui* during 16 weeks of composting. The compost prepared from teak forest litter exhibited the highest population density of the three earthworms, followed by wheat straw compost, mocha litter, soybean straw, palas litter and mustard straw compost. The variability in food resources is an important factor for variation in earthworm population (Fragoso and Lavelle, 1987; Borges and Alfaro, 1997). Among the three epigeal earthworms, *E. fetida*

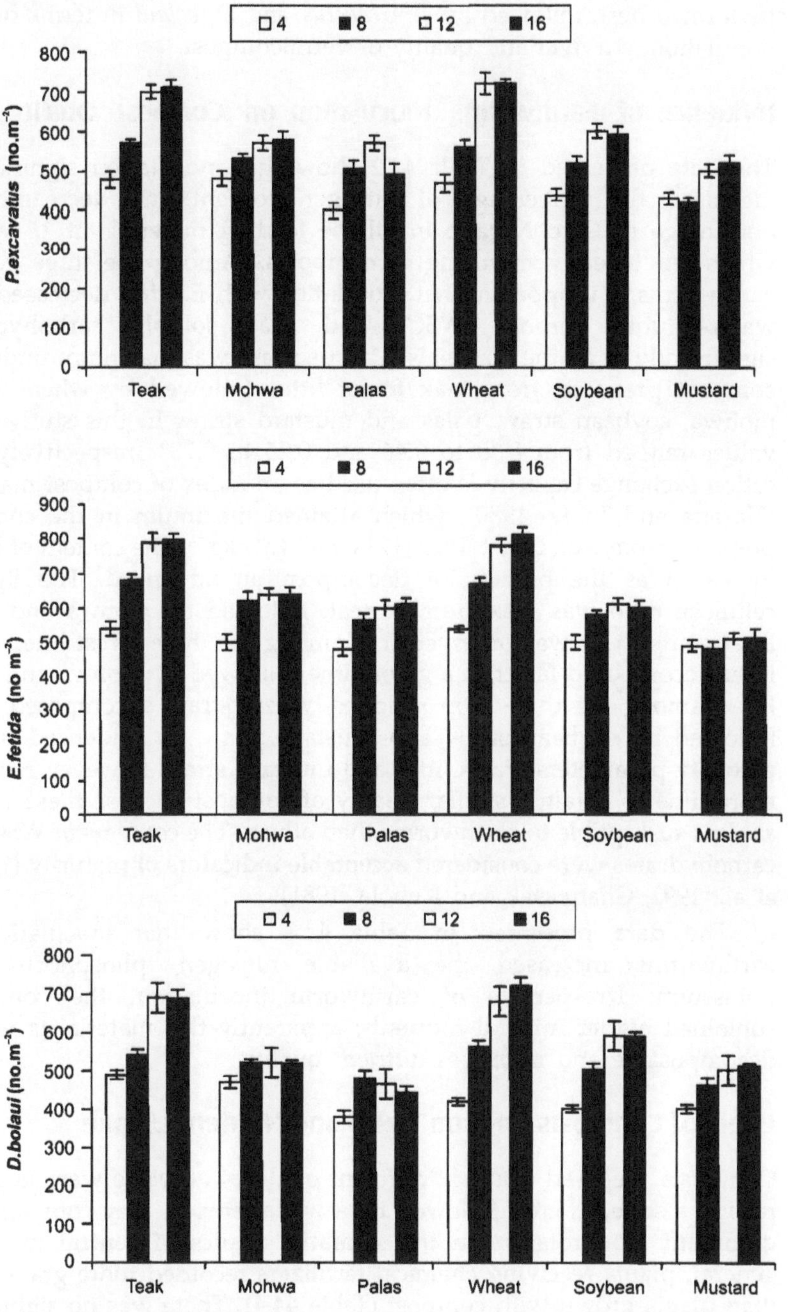

Fig. 44.1 Effect of different food materials on population of (a) *P. excavatus*, (b) *E. fetida* (c) *D. bolaui* at 4,8,12 and 16 weeks after decomposition.

performed best, followed by *P. excavatus* and *D. bolaui* in terms of total population, survival and quality of vermicompost.

Influence of Earthworms Inoculation on Compost Quality

The data presented in Table 44.2 show that inoculation significantly increased the percentage of nitrogen content and decreased the percentage of C to N ratio in all the feeding materials at 16 weeks, which led to early maturing of composts. Among the three epigeal earthworms, it is apparent that inoculation with *E. fetida* decreased both water-soluble carbon (WSC) and water-soluble carbohydrates significantly in all the materials. This decrease was maximum under the compost prepared from teak forest litter followed by wheat straw, mohwa, soybean straw, palas and mustard straw. In this study these values ranged from 0.30 to 0.46 and 0.25 to 0.78% respectively. The cation exchange capacity is often used as an index of compost maturity (Harada and Inoko 1980), which attained maximum in the compost prepared from teak forest litter (121 c mol (p^+) kg^{-1}. The content of lignin increased as the process of decomposition advanced. The lignin/cellulose ratio was maximum in teak followed by mohwa and palas litter compost. It was observed that among the three forest litters, teak litter decomposed faster in a given time, followed by mohwa and palas litter; among the three crop residues, wheat straw decomposed early, followed by soybean straw and mustard straw, as evidenced by the maturity parameters. Palas litter and mustard straw, however, required more time to obtain a similar quality of compost because these wastes are less susceptible to earthworms than others. The contents of WSC and carbohydrates were considered acceptable indicators of maturity (Garcia et al., 1992; Chanyasak and Kubota, 1981).

The data presented in Table 44.3 show that inoculation of earthworms increased the available nitrogen, phosphorus and potassium. Irrespective of earthworm inoculation, the composts contained higher mineral nutrients; apparently this material is readily decomposable and improves nutrient quality.

Effect of Composts on Crop Yields and Nutrient Uptake

Composts prepared from six different qualities of waste were used for raising a soybean crop followed by a wheat crop, a very common and dominant crop rotation in the semiarid tropics of central India. In general, plants receiving chemical fertilizers recorded more grain yield than plants grown with compost (Table 44.4). There was no significant variation in grain yield of soybean with chemical fertilizer compared to

Table 44.2 Characteristics of maturity index of composts prepared from different wastes as influenced by three species of epigeal earthworms (after 16 weeks of decomposition)

Earthworm effect on waste decomposition	*Maturity indices in vermicompost*									
	TOC (%)	*TN (%)*	*C/N Ratio*	*Water-soluble carbon (%)*	*Water-soluble carbo-hydrates (%)*	*CEC (cmol (P+) kg^{-1}*	*CEC/TOC*	*Lignin (%)*	*Cell-ulose (%)*	*Lignin /cell-ulose*
Tectona grandis (Teak) litter compost										
E. fetida	18	1.9	9.5	0.30	0.25	121	6.7	45	8	5.7
P. excavatus	20	1.8	11.1	0.35	0.38	111	5.6	43	9	4.7
D. bolaui	24	1.6	15.1	0.35	0.38	101	4.2	43	11	3.9
Madhuka indica (Mohwa) litter compost										
E. fetida	20	1.6	12.5	0.38	0.30	115	5.7	39	6	6.5
P. excavatus	22	1.6	13.8	0.45	0.46	111	5.0	35	12	2.9
D. bolaui	28	1.5	18.7	0.44	0.48	98	3.5	34	14	2.4
Butea monosperma (Palas) litter compost										
E. fetida	24	1.5	16.1	0.40	0.35	112	4.7	36	8	4.5
P. excavatus	24	1.4	17.1	0.45	0.47	101	4.2	34	12	2.8
D. bolaui	29	1.3	22.3	0.45	0.48	98	3.4	31	14	2.2
Triticum aestivum (Wheat) straw compost										
E. fetida	17	1.7	10.0	0.31	0.28	118	6.9	40	7	5.7
P. excavatus	19	1.6	11.8	0.34	0.34	112	5.9	38	12	3.1
D. bolaui	18	1.5	12.0	0.36	0.39	98	5.4	37	16	2.3
Glycine max (L.) Merr. (Soybean) straw compost										
E. fetida	21	2.0	10.5	0.40	0.35	114	5.4	37	8	4.6
P. excavatus	21	1.9	11.1	0.41	0.46	98	4.7	34	12	2.8
D. bolaui	22	2.1	10.5	0.45	0.48	87	3.9	32	19	1.7
Brassica juncea L. (Mustard) straw compost										
E. fetida	25	1.4	17.8	0.43	0.45	89	3.6	30	9	3.3
P. excavatus	25	1.4	17.9	0.45	0.65	87	3.5	29	10	2.9
D. bolaui	30	1.2	25.0	0.46	0.78	78	2.6	29	13	2.2
LSD at 5%	1.31	0.05		0.022	0.27	9.51		1.9	2.32	

Table 44.3 Effect of three species of epigeal earthworms on available N, P and K status under composts prepared from different wastes[a]

Compost earthworm	NH_4-N	NO_3-N	*Available P*	*Available K*
		mg/kg		
T. grandis (Teak) litter compost				
E. fetida	67 ± 3.4	102 ± 7.1	43 ± 3.2	230 ± 3.1
P.excavatus	60 ± 2.7	96 ± 6.4	30 ± 4.1	231 ± 3.5
D.bolaui	60 ± 2.2	101 ± 5.7	31 ± 2.3	230 ± 4.2
M. indica (Mahwa) litter compost				
E. fetida	53 ± 3.2	79 ± 3.7	35 ± 1.9	202 ± 4.1
P. excavatus	50 ± 2.7	77 ± 2.3	26 ± 2.3	209 ± 3.7
D. bolaui	51 ± 2.1	76 ± 3.1	29 ± 2.2	203 ± 4.2
B. monosperma (Palas) litter compost				
E. fetida	50 ± 4.1	74 ± 3.9	29 ± 2.5	201 ± 5.7
P. excavatus	45 ± 2.8	63 ± 2.6	20 ± 2.5	198 ± 4.7
D. bolaui	42 ± 3.7	60 ± 3.7	19 ± 2.9	198 ± 3.1
Triticum aestivum (Wheat) straw compost				
E. fetida	60 ± 5.1	78 ± 3.9	34 ± 1.5	231± 5.7
P. excavatus	43 ± 4.8	69 ± 4.6	31 ± 2.5	230 ± 3.7
D. bolaui	43 ± 4.8	69 ± 4.6	30 ± 3.5	230 ± 4.7
Glycine max (L.) Merri. (Soybean) straw compos				
E. fetida	60 ± 2.1	88 ± 3.9	33 ± 1.5	238 ± 5.7
P. excavatus	53 ± 3.8	79 ± 2.6	32 ± 2.5	220 ± 2.7
D. bolaui	53 ± 2.8	79 ± 3.6	32 ± 1.5	221 ± 3.7
Brassica juncea (Mustard) straw compost				
E. fetida	39 ± 5.1	58 ± 3.9	29 ± 1.5	198 ± 3.7
P. excavatus	33 ± 2.8	60 ± 4.6	21 ± 2.5	189 ± 6.7
D. bolaui	33 ± 3.8	62 ± 4.6	20 ± 3.5	189 ± 3.7

[a]Estimates of available nutrient status based on dry weight of material and values given the mean of three replications, ± S.E.

composts. Among composts, teak litter recorded a significantly higher grain yield of soybean with the least observed under mustard straw compost. Nitrogen and P uptake were also higher in plots receiving chemical fertilizers and teak litter compost. This compost provided favourable soil conditions for soybean with respect to soil biological and chemical properties that could increase nutrient uptake and provide yields equivalent to those of chemical fertilizers. The residual effect of compost did not contribute more to wheat grain yield due to less

Table 44.4 Effect of composts and inorganic fertilizer on grain yields and uptake of N and P under soybean-wheat system

Compost earthworm	*Grain yield ($mg\ ha^{-1}$)*		*Uptake of N ($kg\ ha^{-1}$)*		*Uptake of P ($kg\ ha^{-1}$)*	
	Soybean	*Wheat*	*Soybean*	*Wheat*	*Soybean*	*Wheat*
T.grandis (Teak) litter compost						
E. fetida	2.3	3.9	138.4	51.9	10.4	8.6
P. excavatus	2.1	3.2	125.8	43.3	9.5	6.7
D. bolaui	2.0	3.0	123.8	40.6	9.1	6.6
M. indica (Mahwa) litter compost						
E. fetida	2.2	3.5	135.1	47.3	9.9	7.7
P. excavatus	1.9	3.2	117.1	43.2	8.5	7.1
D. bolaui	1.9	3.0	117.1	40.5	8.5	6.6
B. monosperma (Palas) litter compost						
E. fetida	1.8	3.0	110.9	40.5	8.1	6.6
P. excavatus	1.7	3.0	104.7	40.7	7.6	6.6
D. bolaui	1.8	3.1	110.9	41.9	8.1	7.1
Triticum aestivum (Wheat) straw compost						
E. fetida	2.3	3.8	141.5	51.3	10.3	8.7
P. excavatus	2.1	3.8	129.2	51.1	9.4	8.7
D. bolaui	2.0	3.7	123.8	49.9	9.1	8.2
Glycine max (L.) Merri. (Soybean) straw compost						
E. fetida	2.1	3.8	129.2	51.3	9.4	8.7
P. excavatus	1.9	3.6	116.9	48.9	8.5	8.3
D. bolaui	2.0	3.6	123.9	47.8	9.3	8.3
Brassica juncea (Mustard) straw compost						
E. fetida	1.6	3.0	98.9	40.5	7.4	6.6
P. excavatus	1.6	2.9	98.7	39.9	7.4	6.2
D. bolaui	1.5	2.8	97.4	38.1	6.7	6.2
N: P: K*	2.5	4.6	152.3	62.3	11.25	10.6
LSD at 5%	NS	0.25	3.61	3.51	1.12	0.87

N:P:K* applied for soybean 25 : 26.2 :16.7 and for wheat 120 : 26.2 : 33.3 kg ha^{-1}.

available nutrients remaining in the soil after soybean harvest. Application of recommended doses of chemical fertilizers on wheat gave significantly higher grain yield over compost treated plots.

Effect of Composts on Soil Biological Activities

Application of compost significantly ($p < 0.05$) increased soil respiration, soil microbial biomass carbon and dehydrogenase activity compared to

Table 44.5 Effect of composts and chemical fertilizers on soil chemical and biological activities of post-harvest soil[a]

Compost earthworm	*Organic C (%)*	*Available N (kg ha⁻¹)*	*Avail. P (kg ha⁻¹)*	*Avail.K (kg ha⁻¹)*	*SMBC (mg kg⁻¹)*	*DHA (ug TPF g⁻¹ 24h)*	*Soil respiration (mg 100g⁻¹ 10d⁻¹)*
T.grandis (Teak) litter) compost							
E. fetida	5.2	170	6.8	249	249	44.7	25.3
P. excavatus	5.1	168	6.3	244	235	41.3	21.3
D. bolaui	5.0	169	6.7	245	245	54.3	20.5
M.indica (Mohwa) litter compost							
E. fetida	4.9	159	6.9	245	179	42.9	22.3
P. excavatus	4.9	151	6.7	241	178	40.9	19.5
D. bolaui	4.9	157	5.9	241	168	40.1	20.5
B.monosperma (Palas) litter compost							
E. fetida	4.8	154	5.5	245	165	41.2	19.8
P. excavatus	4.8	154	6.1	244	157	38.9	16.9
D. bolaui	4.8	151	5.8	241	161	40.1	17.8
Triticum aestivum (Wheat) straw compost							
E. fetida	4.8	154	5.5	245	165	41.2	19.8
P. excavatus	4.8	154	6.1	244	157	38.9	16.9
D. bolaui	4.8	151	5.8	241	161	40.1	17.8
Glycine max (L.) Merri. *(Soybean) straw compost*							
E. fetida	4.8	154	5.5	245	165	41.2	19.8
P. excavatus	4.8	154	6.1	244	157	38.9	16.9
D. bolaui	4.8	151	5.8	241	161	40.1	17.8
Brassica juncea (Mustard) straw compost							
E. fetida	4.8	154	5.5	245	165	41.2	19.8
P. excavatus	4.8	154	6.1	244	157	38.9	16.9
D. bolaui	4.8	151	5.8	241	161	40.1	17.8
NPK	4.8	179	6.9	244	140	40.3	16.1
Initial	4.8	139	5.8	244	140	40.1	16.2
LSD (p = 0.05)	NS	NS	NS	NS	3.41	2.19	1.57

N:P:K applied for soybean 25: 26.2 :16.7 and for wheat 120:26.2 :33.3 kg ha^{-1}.

[a]Estimation of soil chemical parameters, soil microbial activities and activity of dehydrogenase status made on soil dry-weight basis and values given pooled means of three replications.

their initial value (Table 44.5). The magnitude of increase in these characters was less in plots where only chemical fertilizers were applied. In fact, the increase in soil microbial biomass carbon with application of composts was almost twofold more than with inorganic, due to the presence of readily water-soluble carbon, which acts as a source of energy for soil organisms.

CONCLUSION

It is clear from the present study that wastes in excess on cultivated arable farm, or an accumulation of litter not properly managed in forests, if utilized effectively in the form of compost, could improve their quality through Vermicompost in the shortest possible time. This in turn would provide nutrition to plants, improve biological activities of soil and ultimately sustain crop production. Irrespective of organics, the growth performance and population of *E. fetida* earthworm in the vermicompost system were better than those of *P. excavatus* and *D. bolaui*. The study further suggest that to boost the total productivity of crops as well as the system as a whole and to improve soil quality, application of organics through composts may supplement some quantity of inorganic fertilizers.

References

Bremner, J.M. 1965. Organic forms of nitrogen, pp.1179-1237. In: C.A. Black (ed.). Methods of Soil Analysis. Amer. Soc. Agron. Madison, WI USA.

Borges, S. and Alfaro, M. 1997. The earthworms of Bano De Oro, Luquillo experimental forest, Puerto Rico. Soil Biol. Biochem. 29: 231-234.

Brink, R.H., Dyban, P. and Lynch D. Ln. 1960. Measurement of carbohydrates in soil hydrolysates with anthrone. Soil Sci. 89: 157-166.

Casida, L.E., Klein, D.P. and Santora, T. 1964. Soil dehydrogenase activity. Soil Sci. 93: 371-376.

Chanyasak, V. and Kubota, H. 1981. Carbon/organic nitrogen ratio in water extract as measure of composting degradation. J. Ferment. Technol. 59: 215-219.

Edwards, C.A. and Heath, G.W. 1975. Studies in leaf litter breakdown III. The influence of leaf age. Pedobiologia 13: 301-326.

Fragoso, C. and Lavelle, P. 1987. The earthworm community of a Mexican tropical main forest (Chajul, Chiaspas), pp. 281-295. In: A.M. Bonvicini Paliai and P. Omodeo, (eds.). On Earthworms. Mucchi Modena, Maxico.

Garcia, C., Hernandex, T., Costa, F. and Ayas, M. 1992. Evaluation of the maturity of municipal waste composting using simple chemical parameters. Comm. Soil Sci. Plant Anal. 23: 1501-1512.

Hanway, E.D. and Heidal, H. 1952. Soil Analysis Methods used in Iowa State College, vol. 57, pp. 1-31. Soil Testing Laboratory, Iowa State College, Cadar Falls, IA USA.

Harada, Y. and Inoko, A. 1980. The measurement of the cation exchange capacity of composts for estimation of the degree of maturity. Soil Sci. Plant Nutr. 26: 127-134.

Lax, A., Roig, A. and Costa, F. 1986. A method for determining the cation exchange capacity in organic materials. Plant and Soil 94: 349-355.

Lee, K.E. 1983. Earthworms of tropical regions: some aspects of their ecology and relationships with soils, pp. 179-193. In: J.E. Satchell, (ed.). Earthworms Ecology: From Darwin to Vermiculture. Chapman and Hall, London.

Lee, K.E. 1985. Earthworms, Their Ecology and Relationships with Soils and Land Use. Academic Press, Baton Rouge, FL USA.

Meentmeyer, V. 1978. Microclimate and lignin control of hardwood leaf decomposition dynamics. Oecologia 58: 281-285.

Olsen S.A., Cole, C.V., Watanabe, F.S. and Dean, L.A. 1954. Estimation of available phosphorus in soils by extraction with sodium bicarbonate. U.S. Dept. of Agric., Circular No. 939, Washington DC.

Rowland, A.P and Robert, J.D.1994. Lignin and cellulose fraction in decomposition studies using acid-detergent fibre methods. Comm. Soil Sci. Plant Anal. 25 (3 & 4): 269-277.

Satchell, J.E. 1967. Lubricidae, pp. 259-322. In: A. Burges and F. Raw (eds.). Soil Biology. Academic Press, London.

Snedecor, G.W. and Cochran, W.G. 1967. Statistical Methods, Oxford University Press Kolkata, West Bengal, India.

Vance, E.D., Brookes, P.C. and Jenkinson, D.S. 1987. An extraction method for measuring soil microbial biomass C. Soil Biol. Biochem.19 (6): 703-707.

Walkley A. and Black, C.A. 1934. An estimation of the Degtjareff method for determining soil organic matter and proposed modification of chromic acid titration method. Soil Sci. 37: 29-38.

45

APPLICATION OF ORIGINAL ANALOGUES OF PHYTOHORMONES FOR INCREASING PRODUCTIVITY OF HILLED CULTURES (POTATOES, SUGAR-BEET)

Alfonsas Merkys
Institute of Botany Zaliuju ezeru 49, LT-2021 Vilnius, Lithuania

The problem of plant growth regulation and productivity was solved in a novel way and analogues of phytohormones of a highly specific effect created. The most active compounds TA-12, LN-14 and LN-16 were applied for growth regulation of various organs of potatoes and sugar-beet, for modification of growth and organogenesis (potatoes), thus increasing productivity of the cultures. From the economical point of view, application of compound TA-12 (1-2 kg ha^{-1}) increases potato yield (approx. by 6.0 t ha^{-1}), starch content, sugar-beet yield (approx. by 3.0 t ha^{-1}) and sucrose extra content (0.5-1.1%). Compounds TA-59 (0.5 kg ha^{-1}), LN-14 and LN-16 (1.0 kg ha^{-1}) are recommended for use in potato tuber farming. Compound TA-12 as a regulator of potato yield was studied in Latvia, Estonia, Russia, Belarus, the Ukraine and Moldova and demonstrated positive results. Of great significance is the fact that TA-12 increases not only sugar-beet root yield, but also sucrose content. According to the main toxicometric parameter, compound TA-12 is relatively non-toxic and can be applied in agriculture.

46

RED LEAVES CAN INCREASE PRODUCTIVITY OF HORTICULTURAL PLANTS

Vasil A. Zayats
Uzhgorod National University, Department of Entomology and Agronomy, Biological Faculty, Voloshin str. 46, Uzhgorod 88000, Ukraine
e-mail: sharga@univ.uzhgorod.ua

Red leaves may increase the productivity of plants in horticulture due to the involvement of antocyanes in photosynthesis. More often red colour is inherited as a dominant or not completely dominant trait. For red peaches the scheme of Gregor Mendel was approved. In F_1 all peaches were red, in F_2 the ratio of green plants to red plants obtained was 1:3 and by genotype 1:2.1. The spectral characteristic of red plants supporting the idea is that antocyanes enhanced light absorption at wavelength 380-370 nm. From analysis of data for cultivars Red Haven (green leaf) and Negus (red leaf), the absorption of latter is much stronger at light waves 400-470. Negus showed a 42.5% increase in absorption at 430 nm compared to Red Haven. Light absorption intensity in red leaf exceeded absorption in green leaf by 8-10% at 400 – 710 nm.

Development of more productive red cultivars from this plant material is expected in the near future. The cultivar Negus (with dark red leaves) is presently under field evaluation. It seems quite possible that 'red leaf genes' will improve the productivity of other horticultural plants also.

47

INFLUENCE OF ORGANIC SUBSTRATA COMPOSITION ON DEVELOPMENT OF TOMATO ROOTS

I.P. Kozlovskaya
Belarusian State Agricultural Academy, Belarus
e-mail: irinak@www.gorki.unibel.by & mih-mihailov@email.com

INTRODUCTION

Growth of plants largely depends on the wholesome development of their roots. The physiological role of roots is especially significant for plants grown in a protected soil, which have high rates of growth and development, prolonged growing season and consume a great amount of mineral feed elements.

When vegetable cultures are cultivated in containers of small volume, the plant roots are placed in a special substratum, into which nutrient materials are delivered in a dissolved form through a drop bottle. Naturally, such cultivation technology requires a substratum of high quality: the ratio of firm, fluid and gaseous components should be stable throughout the vegetative period. Good aeration of the roots and sufficient quantity of moisture are also required. The acidity value of the substratum should be within the limits 6.2–6.5 and salt content should not exceed 3 g L^{-1} in the dry residue or 4 mS with respect to electrical conductivity.

There are substrata of natural and synthetic origin which meet the indicated requirements. Organic substrata produced from peat moss and synthetic mineral cotton substratum are the most common.

Mineral wool is a synthetic fibre, produced by melting various mineral rocks. The complicated technological process of mineral wool production stipulates its high price; it is not produced in Belarus and is purchased for hard currency. Apart from the high cost, acquisition of mineral wool entails considerable transport and customs expenses. Thus the costs of mineral wool are much higher than for organic substrata. The service life of mineral wool is one, maximum two years. A usedup substratum cannot be utilized and hence usage of it by large hothouse centres creates a serious ecological problem.

From the ecological and economical point of view, organic nutrient substrata have an indisputable advantage. They are produced from a natural stuff, namely, peat moss. Specificity of its physicochemical and biological nature has conditioned broad application of peat as a nutrient substratum in greenhouses.

In olericulture on protected soil, preferably peat moss with a degree of decomposition no more than 15%, which corresponds to 3 according to the Post scale, is permissible for preparing peat nutrient substrata. Such a degree of decomposition is characteristic of peat. The peat is formed in conditions of low microbiological activity; it is acidic and almost sterile, differing in the small contents of feed elements. Preparing organic nutrient substrata, peat moss is moistened and, in conditions of greenhouses with heightened temperature and humidity, its organic matter actively mineralized, which results in deterioration of the physical characteristics of the substratum, disturbance of aeration and feed conditions. Stabilization of substratum properties can be achieved by the addition of organic components to peat.

In our republic, the possibility of using flax awn chaff as a component of peat has been extensively studied in the last few years (Autko et al., 2001; Kozlovskaya, 2001a, b); the influence of substratum structure on development of tomato roots analysed and estimated; and those substrata structures ensuring the most active growth of roots at early stages of plant development determined.

MATERIAL AND METHODS

Moistened peat in a given ratio was mixed with flax awn chaff and mineral fertilizers. The obtained substratum was placed in opaque plastic packets of 24 L volume and used for vegetative experiments (Table 47.1). As under normal production conditions, two tomato plants

Table 47.1 Experimental Schemes

Experiment variant	*Substrata composition*
1 (control)	Peat moss (100%)
2	Peat moss (80%) + flax awn chaff (20%)
3	Peat moss (65%) + flax awn chaff (35%)
4	Peat moss (50%) + flax awn chaff (50%)
5	Flax awn chaff 100%

with a well-shaped first cluster were planted in each packet (Autko et al., 2001).

The volume of roots and volume of active tomato roots were determined at regular intervals (7 days). For this purpose, the plant was taken out of the packet, washed, and the roots dissevered. To determine the volume of roots, plants roots were completely immersed in a graduated measuring vessel filled with water up to a defined value. After extraction of the roots, the volume of superseded water was measured. The volume of the active roots was determined by placing the roots in a solution of methylene blue. The coloured active roots were rewived and the volume of uncoloured inactive roots measured. The experiment was replicated thrice and the outcomes statistically analysed (Autko et al., 2001; Kozlovskaya, 2001a; b).

Already to the first term of observation, the tomato roots grown on different substrata differed essentially in volume (Table 47.2), these differences becoming more pronounced in the subsequent stages of observation. Plants on substrata with the flax awn chaff component had roots of greater volume than plants grown on a peat substratum. At the last examination, the greatest volume of tomato roots was found in plants grown on a substratum with 35% flax awn chaff. The roots volume of plants used on a substratum containing 50% flax awn chaff was slightly smaller. When the flax awn chaff component constituted 20%, the volume of tomato roots was bigger than on control, but less

Table 47.2 Tomato root volume (cm^3) on substrata of different composition

Experimental variant	*Observation terms*			
	1	*2*	*3*	*4*
1 (Control)	15.8	20.0	26.5	34.6
2	19.0	22.3	30.2	40.0
3	17.3	22.2	33.2	45.7
4	16.6	20.6	31.8	42.6
5	15.9	18.9	22.0	27.3

INDEX